QUESTIONS & ANSWERS

Amateur Radio

F. C. Judd (G2BCX)

Newnes Technical Books

Published by Newnes Technical Books
an imprint of Newnes Books,
a division of The Hamlyn Publishing Group Limited,
Bridge House, 69 London Road,
Twickenham, Middlesex TW1 3SB
and distributed for them by
Hamlyn Distribution Services Limited,
Rushden, Northants, England

First published 1980
Reprinted 1982
Revised edition 1986

© Copyright Newnes Books 1986

All rights reserved. No part of this publication may be
reproduced or transmitted in any form or by any means,
including photocopying and recording, without the written
permission of the copyright holder, application for which
should be addressed to the Publisher. Such written
permission must also be obtained before any part of this
publication is stored in a retrieval system of any nature.

ISBN 0 408 00439 8

Printed in England by Whitstable Litho Ltd

PREFACE

Since this book was first published in 1980, the entire Amateur Radio licence and its schedule has been changed. New amateur frequency bands have been allocated and new operating conditions introduced. Many of these changes have proved advantageous so newcomers will benefit accordingly.

This new edition of *Q & A on Amateur Radio* contains full details of the updated licence schedule, new frequency bands and operating conditions etc.

There is also an extra section containing more about radio wave propagation and its close relationship with the '11 year sun-spot cycle'. This is now (1985) at its minimum which will prevail for another year, possibly two. The effect of this is that long distance radio communication is now at a minimum and will not change until the 'minimum of the cycle' has passed and sun-spot activity begins to increase toward its maximum in another 5 or 6 years. This situation applies mainly to the higher frequency bands, e.g. above about 10 MHz, but leaves plenty of 'DX' (long-distance communication) available. Those who take up amateur radio now will have the advantage, in a few years time, of relatively easy-to-make contacts with other radio amateurs in countries all over the world. Meantime there is plenty of opportunity for making long-distance v.h.f. contacts via amateur radio relay satellites which offer some exciting possibilities in modern communications technology.

Computers too are playing a part in amateur radio and there is considerable scope in other fields such as radio teletype, fast and slow scan television and more recently radio astronomy. With an amateur radio licence and the right equipment, the world is at your finger tips.

The generic term *amateur radio* does not quite reveal the true nature of what is primarily a hobby although this does raise what must be the first question in this book, *what is amateur radio?* The basic answer is: two-way communication by radio with recognised amateur status. Although in one sense it is a self-participating hobby, it is one that would be non-existent without other participants.

Normally amateur radio is an indoor hobby because most activity is carried out at home but there is also outdoor activity. For example, one can have a mobile station which permits two-way communication from a car and can also operate a portable station. Amateur radio can be taken on holiday anywhere in the UK and in other countries as well

Amateur radio clubs and societies regularly hold field days in which groups set up portable stations, usually in contest with each other and there are mobile rallies which provide some 'family' participation and frequently the opportunity to purchase radio equipment at low prices.

Although most amateurs now buy transmitters etc. ready to use there is still much scope for 'home construction'. So we have a hobby that provides a wide range of interest and activity and allows world wide contact with others having the same interest. There are still areas to explore and discoveries to be made despite the tremendous advances in radio and electronics technology.

This book cannot be regarded as a 'textbook' of all the technicalities involved in amateur radio equipment, circuitry, transmission modes, aerials (antennas), measuring methods, mathematics used in radio and electronics etc. Titles of textbooks and magazines devoted to amateur radio are therefore mentioned in the text and listed at the end.

Cantley, Norfolk   F. C. Judd G2BCX

CONTENTS

1
Introduction to Amateur Radio 1

2
The Radio Amateurs' Examination and Transmitting Licence 16

3
Amateur Radio Technology 31

4
Equipment for an Amateur Radio Station 58

5
Aerials for Amateur Radio Transmitting Stations 67

6
Operating Procedure, Signals and Codes 88

7
More about Radio Wave Propagation 104

Appendix: Useful Information 116

Index 121

The author's QSL card

1
INTRODUCTION TO AMATEUR RADIO

What is the origin of amateur radio?

Almost three quarters of a century ago Marconi made the first 'wireless' transmissions which particularly aroused the interest of a few scientifically minded experimenters, who in the early 1900s, formed groups around the country. The London Wireless Club (circa 1913) was one and it applied for what was the first officially issued amateur radio transmitting licence. This was obtained from the Post Office, which also became the controlling authority for all radio transmissions and continued to be until a few years ago.

The enthusiasts set up their transmitters and receivers which, including component parts, had to be 'home constructed' and then proceeded to experiment either from their own ideas or from what little technical information could be passed from one to the other 'over the air'. The need for a national organisation, as opposed to localised clubs, for collection and exchange of ideas and information was therefore becoming necessary. The experimenters soon proved that, with the long wavelengths allocated to them, considerable distances could be covered with quite acceptable reliability. However, this attracted the attention of commercial and broadcasting organisations and the experimental fraternity had eventually to relinquish these wavelengths. They were left with only the short wavelengths (below 200 metres) to work with, largely because these had been considered as useless except for transmission over very short distances. In fact it was doubted whether the experimenters would get signals to travel much further than their own back gardens! Much to everyone's surprise however and given certain conditions, communication on short waves was shown to be possible over hundreds and, as was discovered later, thousands of miles.

By now experimenters were beginning to regard themselves as bona fide amateurs in this field and decided it was time they had a national body to protect their interests. The *Radio Society of Great Britain* was formed in the UK, *The American Radio Relay League* in the USA and similar organisations in other countries; later these became affiliated to an international organisation: the *International Amateur Radio Union*. Even so, commercialism and also broadcasting, both with world wide interests, once again stepped in and, although the amateurs virtually lost the short waves as well, they were at least left with an allocation of specified narrow bands within the total span of the short wave spectrum from about 200 down to 10 metres. Dedicated radio amateurs are never deterred and with the help of their national societies and international body have succeeded in retaining, except for a few changes, most of the original amateur wavelength allocations to the present time. New bands have quite recently been allocated.

How has amateur radio developed?

Over the years amateur research and investigation into many different areas of radio communication has resulted in a number of discoveries, many of which have been adopted by 'professional' concerns. The amateur tradition of building one's own equipment and the inclination to experiment still prevails, although perhaps to a lesser extent, which may be largely due to the availability of ready to use transmitters, receivers and auxiliary items and even aerials as well. Communication as such is still world wide but has become a slightly more social function in that exchanges of information, or to put it less formally, conversation over the air, is frequently more topical and generalised, rather than wholly technical.

Because technological advancement has been very rapid, especially over recent years, it could in fact be said that the 'technology boot' is now on the other foot with the amateurs emulating the professionals, using their ideas and developments. After all, the pace at which technology has advanced could only have been set and maintained by professional concerns with their vast resources for research, design and development. The radio amateurs can devote only their spare time and limited amount of money to their interest. Even so they are not lacking in technology and today have their own communications satellites orbiting the earth and have transmitted from one country to another by bouncing radio signals off the moon. They have also quite sophisticated repeater stations for increasing working ranges for mobile transmitters operating at very high frequencies and they are also very up to date with picture (facsimile and television) and teleprinter transmission with communication being carried out over very long distances.

Are there many radio amateurs?

In an average sized town there could be a dozen or more active radio amateurs. In any large city such as London or Birmingham and their suburbs there may be several hundred. A single county might easily have the same number or even more. The total number of radio amateurs in the UK, including the Channel Islands, is more than 50,000 and the world total may even have exceeded 1 000 000. You will certainly not be alone!

How do radio amateurs identify themselves?

Each radio station is assigned its own callsign with a prefix to indicate the country (and there are about 200 countries) in which the station is licensed. The prefix is a combination of letters and figures, or both, that precede the station callsign but identifies the country, e.g. G is for Great Britain, GW for Wales (counted as a separate country), KH6 is the Hawaiian

Islands and so on. Large countries such as the USA and Canada also have Province or area numbers, e.g. VE denotes Canada but VE1 signifies the Canadian Maritime Provinces and VE2 the Province of Quebec (amateur radio improves your geography). However, the various licensing authorities in different countries have to keep track of the amateur radio stations so each station is given a callsign which consists of the country/province prefix and the station identification figures and/or letters, or combination of both. For example, in G2BCX (the author's callsign) G is for England and 2BCX is the station identification. The callsign PA3AEN is that of a Dutch station in which PA denotes the country, the figure 3 a licence classification and AEN the station identification. But more of this in chapter 6.

How active is amateur radio?

Almost irrespective of time, day or night, or season, the amateur radio bands are never idle. One can always be sure of finding a contact somewhere, near or far, on one of the various wavelength allocations and there are a fair number of these to choose from. If the 10 or 20 metre bands are a little quiet, then there may be activity on 80 or 2 metres. Depending on atmospheric and other conditions the amateur bands are always active; indeed they should never be otherwise. Complete occupation is one way of letting the authorities know that radio amateurs are making full and necessary use of their allocations. See also Chapter 7.

What is the meaning of amateur?

An amateur is someone who pursues an interest without being paid to do so and for what it may offer in terms of acquiring greater knowledge of the subject, improving his skill at the art and of course personal satisfaction and enjoyment. If an amateur utilises his expertise for remuneration then in effect he has become professional. It has already been said that many radio amateurs are also qualified engineers, practising in one branch or another of radio, electronics and television. This is their profession, but curiously enough they seem quite able to maintain their amateur radio activity on a strictly amateur basis.

Can anyone become a radio amateur?

There is very little restriction and certainly no discrimination in colour, religion or sex. Applicants for a licence in the UK must normally be of British nationality (but see below), must be 10 years of age or over, and must have passed the radio amateur's examination. Radio amateurs already licensed in the UK can obtain a special licence or permit to operate in another country whilst temporarily resident or on holiday there and, equally, foreign visitors to the UK can obtain a licence to operate provided that they furnish proof of being licensed in their own country. There are however, some limitations and requirements in this respect and information should be obtained from the appropriate licensing authority or national society, for example in the UK, the Department of Trade and Industry and/or the Radio Society of Great Britain.

Is the radio amateurs' examination difficult?

For anyone with average intelligence and a willingness to devote time and self study or special evening classes, the answer is No. The technical knowledge required is largely basic and the examination questions are set out in a much more simplified way than in previous years. Age need be no barrier and many of today's radio amateurs obtained their licences at the previous minimum age of 14 and even at the age of retirement quite a large number have succeeded in passing not only the technical examination but the morse code test as well. In almost every part of the UK where there is a fairly substantial population of radio amateurs, evening classes for the RAE (Radio Amateurs' Examination) are usually run from about September to May, for which a fee is payable. Following this the examination may be taken and there is usually a small extra charge.

Is the morse code necessary?

The 12 words per minute (w.p.m.) morse code test is not required for Class B licence operation on the amateur band frequency allocations above 144 MHz (2 metres). It is required only for the full Class A licence for operation on all the frequency band allocations from 1.8 MHz through to 24 000 MHz. Note however, that since April 1984 Class B licence holders have been allowed to send morse code (c.w.) on what is known as 'all-mode' frequencies within the allocated band, e.g. 144 to 146 MHz and 430 to 440 MHz etc. It is a requirement of this facility that the operator of the station must announce the callsign in 'speech' before and after any morse code transmission. This facility is to allow Class B licence operators to practice morse before applying for the 12 w.p.m. morse test. The facility is subject to review by the Department of Trade and Industry and could be withdrawn if abused.

Is morse code difficult to learn?

Morse code is relatively easy to learn when one is fairly young. The older one becomes the longer it takes to achieve higher speeds of sending and, more so, receiving. Learning the code, i.e. memorising all the dot and dash combinations, is not difficult at any age. There are however no short cuts to learning the code itself or in attaining proficiency. There are alleged methods for working up speed, but it is well known that regular practice is the only sure way of attaining a speed of 12 w.p.m., or higher. Depending on memory and brain reaction all except very elderly persons should be able to reach a speed of 12 w.p.m. within a few months with regular daily practice of 15 to 20 minutes per day. But don't stop at 12 w.p.m., continue for a while until a speed of about 15 w.p.m. is reached before taking the morse code exam. This will ensure a better chance of passing, particularly as the test now costs £15.00. If you should fail it will cost another £15.00 and your travelling expenses to take another test.

Is the amateur radio licence expensive?

By comparison with the cost of a television licence the answer is no. The cost of an amateur radio licence, either Class B or Class A is currently £12.00 and this covers home station as well as mobile (car) and portable operation but not, repeat not, maritime mobile operation from a vessel at sea. A separate licence is required for maritime mobile operation and costs £12.00 per year. There is also a fee payable before the issue of the licence for an authorised officer of the Department of Trade and Industry to inspect the equipment and its installation on the vessel. Note that a Class A licence must be held before a maritime licence will be issued (see also Chapter 2).

Can a station be set up with surplus equipment?

At one time there was plenty of surplus radio equipment available which could be obtained at extremely low prices and modified for amateur radio operation with little difficulty. Unfortunately those days are gone, although one can still find what could be termed surplus transmitters and receivers that have become obsolete for professional purposes, but adequate for amateur use with suitable modifications. Without specific knowledge and certain test instruments the use and modification of surplus equipment may present problems and is not particularly recommended. Building from kits of parts was at one time a means of saving money but nowadays can cost almost as much as the purchase of ready made equipment. Secondhand transmitters, receivers and auxiliary items in good condition are always available, usually at attractive prices, and are frequently advertised in periodicals devoted to amateur radio.

What does it cost to set up a station?

What you could actually get for various amounts of money might be a better way of considering the question. For instance, for about £100 one may be able to obtain a good secondhand but older model v.h.f. (2 metre) transceiver suitable for fixed or mobile operation and for £300 to £400 a new model of similar type. For £300 up to say £500 one might get a second-hand fixed station outfit consisting of a transmitter and receiver (or combined) for the longer wavelength bands, i.e. 80, 40, 20, 15 and 10 metres, with possibly one or two extras such as a microphone and morse key. There is of course no limit to what *could* be spent. The usual practice if cash is a little short is to buy what one can afford to get started and build up from there. Many small auxiliary items can be home built and this can add interest and activity but on the other hand those who have sufficient knowledge and practical experience would find that the construction of equipment for a whole station is not beyond the bounds of possibility.

Does an amateur radio station take up a lot of room?

If one has the available space then a whole room is ideal and a small part can be allowed for constructional work. On the other hand a complete transmitting and receiving station need take up no more room than the equipment itself, which today is very compact. For example, a typical v.h.f. 10 watt output transceiver for the 2 metre band may measure little more than $250 \times 150 \times 75$ mm ($10 \times 6 \times 6$ in). The only other requirement is an aerial which, for the 2 metre band for example, need only be 1 to 1.5 metres (40 to 50 in) long and quite slim. It can be mounted on a chimney stack or even in the loft.

A multiband transmitter with the maximum power output permitted for amateur radio, namely 150 watts, plus a full range communication receiver to cover all the amateur wavelengths, would take up little more than the space needed for a hi-fi system and could therefore be accommodated on a wide bookshelf or small table. Aerials for the longer wavelength bands are of necessity fairly large and while for some bands, like 10 and 15 metres, could be set up in a large loft, they normally require space outdoors, not only to accommodate the size but also for reasons of transmitting efficiency. Many radio amateurs set up their stations in garden sheds which, if heated in the winter, can be quite cosy and one doesn't have to worry too much about tidiness or even of having to stow equipment away out of sight when not in use in order to satisfy the other half or keep it out of reach of children. All radio amateurs refer to the station environment, be it living room or garden shed, as 'the shack' which probably stems from the fact that the vast conglomeration of equipment used by the early experimenters could never really be set up anywhere else except in a garden shed.

What do radio amateurs talk about?

This may depend entirely on a subject raised by one or the other and more so in a group of amateurs in communication; this is usually called a 'net'. Conversation may be topical or technical and about the only subjects prohibited are religion, politics or any form of advertising, or anything of obscene nature. The playing of music is also prohibited but one can record a speech or morse code transmission from another amateur station with permission from that station and replay it over the air with acknowledgement. While conversation with local stations may be topical, the more distant stations (DX contacts) are usually confined to exchanges of signal reports and information concerned with locations, equipment, aerials used, weather conditions etc. and possibly technicalities concerned with any special tests that may be carried out in connection with equipment such as a change of aerial or power.

Can anyone operate the station?

No one except the licence holder or another licensed amateur is allowed to transmit. A visiting licensed radio amateur may operate your station provided that his licence is equal to yours and vice versa and the visitor must also sign your log book at the entries of the transmissions made at the time. Holders of an amateur radio certificate may operate another licenced amateur station.

What is short-wave listening?

Many radio enthusiasts confine their activity to listening only on the short-wave broadcasting and amateur wave bands but frequently do experimental work in connection with receivers and auxiliary equipment and even aerials. They listen mainly for stations at great distances or for what are often referred to as rare stations, particularly amateur stations which may transmit for only a short time from some remote island or part of the world. Rare stations may be those attached to expeditions, for example, at distances that are heard only when atmospheric conditions are very good. Short-wave listeners or SWLs can belong to local clubs and the RSGB issues a recognition number preceded by the initials BRS (British Receiving Station) e.g. BRS 1234, which indicates that when reports from such stations are sent to the transmitting station they come from a recognised and practised listener. Other societies issue a similar form of identification preceded by SWL and many BRS and/or SWL stations have their own reporting or QSL cards on which their identification is prominently displayed. Amateur radio and broadcast stations alike always welcome reports from short-wave listeners, provided they are sensible and contain useful information, and will send a QSL or confirmation of receipt card in return.

What is citizen's band (CB) radio?

This is *not* amateur radio and was first allowed in the UK in November 1981. It is widely practised in many other countries and as it name suggests is a radio communication facility for citizens in general. A common frequency allocation is 27 MHz (approx. 9 metres) which is used in the USA and elsewhere. CB radio has caused problems through abuse, interference to other services, and ill-mannered clashes between users. Properly organised and monitored and used with personal discipline, it does no doubt have its uses, especially in vast areas like the USA, Canada and Australia etc., where communication by other means may be limited, or for emergency purposes in disaster areas in the event of widespread floods or earthquake.

How does CB differ from amateur radio?

CB tends to be treated more as a mobile telephone system, and most radios are car-mounted or hand-held. It is not necessary to pass an examination in order to obtain a licence to transmit. CB is limited to the use of a narrow band of frequencies only and the permitted power is low, so the effective working range is only about 15 km (approx 10 miles) although there are times when this may be extended considerably when what is known as 'Sporadic E' conditions prevail especially during the summer months, mainly June and July (see Chapter 7, More About Radio Wave Propagation).

There are two UK CB radio bands: 27.6 to 27.9 MHz and 934.025 to 934.975 MHz, both bands being for f.m. (frequency modulation) only. Readers interested in taking up CB radio are referred to the Newnes publication *Q & A CB Radio* by this author. Readers are reminded that the use of any unlicensed transmitter could render the operator liable to prosecution and confiscation of the equipment.

Are there any restrictions regarding aerials?

This is an important question and there are certain restrictions concerned with the erection of aerials outdoors. Although indoor aerials, in the loft for instance, present no problems, this is not the ideal place for an aerial as both transmitted and received signals can be greatly reduced in strength and therefore limit the working range. Outdoors and as high as possible is the ideal situation for any radio aerial but planning permission may almost certainly be required by local authorities even for attaching an aerial, however inconspicuous, to a chimney stack or other part of a house. It would be prudent therefore, to make enquiries at your local council office or other appropriate authority.

Assuming that an aerial may be erected somewhere outdoors, on rooftop or mast, then precautions must be taken to ensure that it is safe and cannot fall and cause damage to your own or other peoples' property, or, injury to others. For this reason, and such an occurrence is always possible, it is wise to have insurance or have such an event covered in existing property insurance.

Is there a national society for amateur radio?

In many countries there is a national society and dozens of local clubs and societies, almost to the extent of one in every large town and even one in every large suburban area in widespread cities like London. The national society in the UK is the Radio Society of Great Britain and anyone, anywhere, interested in radio can become a member. The function of the RSGB is to look after the interests of its members, whether licensed amateurs or short-wave listener members and may also represent and speak for British members in negotiations with the national licensing authority and at international conferences concerned with wavelength/frequency allocations etc. The RSGB provides many facilities for members which include a monthly technical journal, *Radio Communication*, popularly known as 'Radcom', regular meeting for members, usually organised by local RSGB groups, contests and field days, achievement certificates, news bulletins by radio, slow morse practice transmissions, beacon radio stations for frequency checking etc., a QSL card service and a technical publications sales department. There are various grades of membership and details of these and all other RSGB activities, facilities and services can be obtained from its headquarters at Lambda House, Cranbourne Road, Potters Bar, Herts EN6 3JW.

Are there local and independent amateur radio clubs?

You are almost certain to find an independent local club or society, or RSGB group, within easy reach of your own location. Nearly all independent clubs are affiliated to the RSGB so your interest in the national society is also catered for and you will of course be able to meet local members of the RSGB. One of the main 'club' facilities is to provide a venue for meetings in connection with their own activities and those of the RSGB if affiliated to it. The 'business side' of meetings is usually short, except for something special and invariably followed by items such as a talk on some technical subject of amateur interest, often by one of the members, or there may be a 'junk' sale of equipment surplus to the requirements of members. Many larger clubs also run contests and mobile rallies as well as purely social functions for members and their families. There are also specialist organisations such as the UK f.m. group for those who have special interest in very high frequency transmission and AMSAT, which is also international and is an organisation concerned with amateur radio communications satellites. So radio amateurs and shortwave listeners alike are well catered for when it comes to society, club and group activity and it is to everyone's advantage to become a member. Details of local clubs and/or societies can be obtained from the RSGB at the above address.

What can one contribute to amateur radio?

Far too many people think in terms of 'what can I get out of it' rather than 'what can I contribute to it'. A very great deal can be got out of any hobby but with amateur radio even greater satisfaction can be obtained from contributing something to it, however small. There is a general code of behaviour which the majority of amateurs practice. Being friendly to others, observing the etiquette of good operating, being helpful to newcomers and not allowing one's hobby to interfere with other duties and so on. Among the radio amateur fraternity are hundreds of blind and physically handicapped operators who frequently need the assistance of others but rarely, if ever, ask for that help. Here then is one small contribution that can be made to the hobby – assistance to others less fortunate than oneself. While on this subject it should be mentioned that blindness or physical handicap need not deter one from becoming a radio amateur; it is an ideal pastime, if only for keeping in direct touch with the world at large.

There are other ways of contributing, such as service to one's local club by assistance with its running, donation of old components and equipment for junk sales to raise club funds and which is also a means of enabling those with limited spare cash to buy parts for constructing some needed item. Articles for club newsletters, the construction of equipment for club use, or help in setting it up, are also valuable contributions.

Are there magazines and books devoted to amateur radio?

Yes — details will be found at the end of the book. Members of the RSGB receive its monthly publication, *Radio Communication*, free of charge but this cannot be purchased on bookstalls or by non-members. There are of course other monthly journals and while not all cater entirely for amateur radio, they do carry technical articles and regular features of interest to amateurs. The RSGB Publications Department has books devoted to numerous subjects of 'radio' interest which can be purchased by non-members. Many local clubs and groups produce newsletters often containing technical articles written by members or specialist contributors. Amateur radio as well as its allied subject of electronics is well catered for with respect to books and magazines.

Where does one start with a view to becoming a radio amateur?

If you decide to take up photography it is but a simple matter to buy a camera and some film and begin taking pictures. Unfortunately it is not quite so easy to take up amateur radio. There is the radio amateur's examination to take and for those starting from scratch it is necessary to decide whether to self study to obtain the necessary knowledge or to take evening classes. The former means obtaining suitable books and strict devotion to spare time study and the latter may mean waiting for the classes to begin around September. Either way takes time. However, you are free to take up shortwave listening requiring neither exams or licence and from this get the feel of radio communication by listening to radio amateurs and also distant broadcast stations. You will also gain some status if you join the RSGB or a local society and this is recommended, because it will enable you to make personal contact with licensed radio amateurs as well as others like yourself, who are just beginning. Remember that virtually all radio amateurs were at one time beginners and the vast majority made the grade via shortwave listening so you will find both friendly assistance and especially sympathy should you be unfortunate enough to fail the examination the first time. To put your mind at rest on this possibility however, note that the number of passes is usually quite high. More will be found about the all important RAE in Chapter 2.

What is RAYNET?

RAYNET is a national emergency communication service, voluntarily provided and operated by licensed radio amateurs, although shortwave listeners can become members and take certain active participation. As RAYNET suggests the full title of this organisation is Radio Amateur Emergency Network, formerly RAEN and it came into being as the result of the flood disaster in 1953 when large numbers of radio amateurs set up emergency communication for ambulance services and the Police. Organisation on a national basis soon followed with full approval by the licensing authority, at that time the Post Office. RAYNET is authorised to operate in conjunction with the Police and with the St. John's and British Red Cross organisations in connection with ambulance services etc. and although it provides nationwide coverage it functions on a county/area basis. Exercises are carried out regularly and often in conjunction with the services it caters for. Since its formation RAYNET has been brought into operation in numerous real emergencies. Here then is another way in which a radio amateur can contribute not only to the hobby but to the community as a whole.

2

THE RADIO AMATEURS' EXAMINATION AND TRANSMITTING LICENCE

What is the Radio Amateurs' Examination?

The Department of Trade and Industry requires that every applicant for an amateur licence, A or B, must have passed the radio amateurs' examination as evidence of possessing the requisite knowledge of theory and the terms of the licence and its requirements. Every applicant for the amateur licence A must also have passed the Post Office morse code test within one year before the time of applying for the licence. The examination consists of a total of 95 multiple choice questions. A multiple choice question has a format similar to the following:

IS THE EARTH — (or may be posed THE EARTH IS —)
(a) Round
(b) Square   *Mark correct answer with a tick.*
(c) A flat plane

How are the questions set?

Questions are normally set to test the recall of facts, comprehension, and application. The simple question above is factual recall whereas a comprehension question is designed to check the understanding of what has been learned. The application question is aimed at verifying both understanding and application of the knowledge to a particular problem. Sample papers with typical questions and answers are available from the City and Guilds of London Institute (address at end of the book).

What other information is available?

The City and Guilds of London Institute will also provide other information concerned with the RAE such as programmes of courses (normally evening classes at an appointed venue) and they recommend that students should if possible carry out practical work to augment the theoretical training, provided of course that any experimental equipment constructed is not allowed to transmit.

Are any other qualifications required?

The selection of students for the course is within the discretion of the college or institute concerned but no specific educational qualifications are required. A candidate for the examination can be accepted whether or not a course has been attended. Obviously the candidate would have the requisite knowledge. The scheme is available outside the UK and information about this is obtainable from the C.G.L.I.

Is a pass certificate awarded?

Certificates are awarded to candidates who pass both sections of the examination and these indicate the examination taken and the *grade* of performance for each candidate which may be 'Distinction', 'Credit', 'Pass' or 'Fail'.

When and where are classes run?

Evening classes for the RAE are normally run once a year beginning about the first or second week in September (earlier in Scotland). They last for about 20 weeks with a class session once a week. Enrolment is made direct to the appointed college or institute and the requisite fee is payable on enrolment. This normally becomes forfeit if the applicant fails to attend the classes.

Most colleges etc. have either a limited number of places for students or do not run the classes if the total number of applicants is below a certain figure. It is as well therefore to apply for a place in good time before the classes are due to begin. For this reason it is not possible to include a list of colleges and other places where classes are held. However, such information can be obtained by writing, shortly before evening classes commence, to the City and Guilds of London Institute, or the Radio Society of Great Britain, or the local Regional Advisory Council for Technological Education.

What are the possibilities of home study?

For anyone with absolutely no knowledge of radio it would be very difficult to absorb all the necessary theoretical and practical information from textbooks, at least within a few months. Those with some basic knowledge to begin with might well succeed, as indeed many have. Advice here, regardless of technical knowledge level in radio, is to attend evening classes. You will be much more assured of success and there is a good deal to know about the transmitting licence itself and the special problems associated with amateur radio which are not fully covered in textbooks.

Is a syllabus available for the RAE course?

Yes and it provides details of all the subject matter related to the examination objectives from which the multiple choice questions are derived. A *full* copy of the syllabus and objectives is available from either the City and Guilds of London Institute or the Department of Trade and Industry (addresses at the end of the book).

What are the general items covered by the syllabus?

The syllabus is in two parts the first being 765-1-01 and covers the conditions of the amateur transmitting licence, transmitter interference, problems concerned with various modes of modulation and frequency checking equipment.

The second part (765-1-02) covers operating practice and procedures, electrical theory, semiconductors, radio receivers, transmitters, radio wave propagation and aerials and functional measurements concerned with the performance of transmitters. It is wise to read the whole syllabus and study the full range of subjects that are dealt with (see also Chapter 3).

How is the radio amateurs' examination arranged?

The sitting for the examination normally takes place where classes are held. Arrive well before time and take pens, pencils and whatever else you may be required to provide. The use of a pocket calculator is allowed.

The examination for 765, Radio Amateurs consists of two separate papers: 765-1-01 Licensing Conditions and Transmitter Interference contains 35 multiple choice questions and 765-1-02, Operating Practices, Procedures and Theory contains 60 multiple choice questions. Questions are allocated to the syllabus sections as follows:

765-1-01 — Licensing conditions and transmitter interference
(1 hour)

	Syllabus	Questions
1.	Licensing conditions	23
2.	Transmitter interference	12
		35

765-1-02 — Operating practices, procedures and theory
(1¾ hours)

	Syllabus	Questions
1.	Operating practices and procedures	5
2.	Electrical theory	11
3.	Semiconductors	9
4.	Radio receivers	9
5.	Transmitters	9
6.	Propagation and aerials	10
7.	Measurement	7
		60

There is a break of about 15 minutes between the two papers.

When does one take the RAE?

Those already possessing the requisite knowledge, or who have attended classes and feel competent to do so, can sit the examination in December. However the examination is held again in May following the winter evening classes and is the more usual time for those who have attended classes. Applications for the December sitting must be made before October 15th and for the May sitting before February 15th. Most colleges have a limited number of places for external candidates (those who have not attended classes) so it is wise to enquire in good time to secure entry.

The results are made known shortly after the examinations and are sent direct to the examination centres who then forward them to the candidates.

Are there any exemptions?

No exemptions are given for any part of the radio amateurs' examination. No other qualifications are accepted for exemption.

What is the success rate for the RAE?

More than 60% of candidates secure a pass although this figure applies to exams held before the end of 1978. The multiple question format may well be the means of generating a higher number of passes per given number of candidates as it is claimed to be somewhat easier than the previous system of specific question and answer with explanation.

When should the morse test be taken?

As success in the morse test remains a valid qualification for twelve months from the date of the test for the purpose of obtaining an amateur licence or an amateur radio certificate, it is advisable to take the morse test after passing the RAE. An applicant for an amateur licence A or an amateur radio certificate who passed a morse test more than twelve months before applying for a licence or certificate would have to pass a further morse test.

Where can the morse tests be taken?

First an application form has to be completed and this is available from the Department of Trade and Industry (address at the end of this book). An application form is also contained in a booklet issued by the Department of Trade and Industry called *How to become a Radio Amateur* issued free on application. The application form must be sent to an allocated test centre listed on the back of the application form. Note: whilst the test will be arranged as far as possible to suit the candidate's convenience, British Telecom International who conduct the tests, cannot guarantee that it will be possible to hold the test at the time stated in the application form.

The examination fee cannot be returned to any candidate who withdraws or fails to attend for the examination, nor can it be transferred from one examination to another at a later date. Special arrangements are made for handicapped persons who are unable to travel to one of the test centres, by telephoning or writing to the nearest test centre.

Is a fee payable for the morse test?

The fee for a morse test is £15.00 and a cheque for that amount must accompany the application to take the test.

What does the morse test consist of?

In 'sending' tests a candidate is required to send 36 words (averaging five letters per word) in three minutes without uncorrected error, not more than four corrections being permitted, and 10 five figure groups in one and a half minutes without uncorrected error, not more than two corrections being permitted.

In the receiving tests, a candidate is required to receive 36 words (averaging five letters per word) in plain language in three minutes and 10 five figure groups in 1½ minutes. Each letter or figure incorrectly received counts as one error. A word in which more than one letter is incorrectly received counts as two errors. More than four errors in plain language and more than two errors in the figure test will result in failure.

The tests will not include punctuation or other symbols. These particulars are summarised as follows.

Type	Length of test	Duration of test	Sending Maximum No. of corrections	Sending Maximum No. of uncorrected errors	Receiving Maximum No. of errors
Plain language	36 words (average 5 letters per word)	3 mins.	4	0	4
Figures	10 groups	1½ mins.	2	0	2

Is tuition in morse code available?

There are several schools at which tuition is given, and particulars may be obtained from the local Education Authority, the City and Guilds of London Institute or the RSGB. Slow morse practice transmissions are made weekly in various parts of the country by volunteer licensed radio amateurs. Details of these can be obtained from the Radio Society of Great Britain.

Can I obtain a copy of the Department of Trade and Industry Amateur Radio Transmitting Licence?

Yes and it is worth applying for since it covers a very wide range of special requirements and conditions some of which will be the subject of questions in the RAE. A full copy of the amateur radio licence (Class A and B) and which will be a 'specimen cancelled copy' can be obtained from:

   The Radio Licensing Unit
   Post Office Headquarters
   Chetwynd House
   Chesterfield
   Derbyshire S49 1PF

What is the fee for the licence?

The fee for either licence Amateur A or Amateur B is £12.00 per annum and the fee is payable annually. Note that the fee may be increased from time to time.

What is the Amateur Licence A?

This licence authorises the use of all allocated amateur bands of frequencies and modes of modulation and is the main amateur transmitting licence. The licence contains details of the operating conditions, the schedule (frequency bands and modulation modes etc.) as well as other requirements such as equipment for frequency checking. In order to qualify for a Class A licence you must satisfy the following requirements:

   (a) be over 10 years of age
   (b) be a British subject
   (c) have passed the RAE
   (d) have passed the British Telecom morse test
   (e) have paid the licence fee.

Three callsign letters are issued in strict alphabetical sequence although the complete callsign will consist of an additional letter and figure known as the prefix. The first letter will denote country and in the case of the UK as a whole, will be G for England, GM for Scotland, GW for Wales, GI for Northern Ireland, GU for Guernsey, GJ for Jersey and GD for the Isle of Man.

What is the Amateur Licence B?

This licence does not authorise the use of frequencies below 144 MHz but it does permit sending messages in morse code but only in the 'all mode' sections of the band(s). The station callsign must be given in 'speech' before and after any morse transmissions are made. In order to use morse in the normal c.w. mode and in the allocated section, the British Telecom morse test must be taken and a Class A licence obtained. No morse test is required for the condition of sending morse, as above, i.e. in the 'all mode' section of the bands but the Department of Trade and Industry reserve the right to withdraw this facility at any time. Three letter callsigns are issued to Class B licence holders with an appropriate prefix as in the previous question.

What is the schedule?

This is a section included in the licence and is reproduced in Fig. 1. It lists the allocated amateur frequency bands for both Class A and B licences and includes the various modes of modulation permitted for the different bands as well as the permitted transmitting power allowed for each band. Note that 'transmitting power' is no longer given as so many watts but instead is rated in 'dBW' or decibel watts. For example '20 dBW' related to a power of 1 watt means that the actual power allowed in this case is 100 times 1 watt or 100 watts.

For the sake of convenience the modes of modulation are given (in the schedule) in plain language. There are however 'coded' versions of the various modes which along with other technical information is included in the licences.

How, when and where may an Amateur Station be operated?

At home of course and in any vehicle (not public transport or aircraft) or in any vessel but not on the sea or within any estuary, dock or harbour unless an Amateur Maritime Mobile Licence is held. Operation is permitted on virtually all inland waterways. The station may be operated 'pedestrian' (portable equipment carried and used while walking about).

The station may be used for the purpose of sending to and receiving from other licenced amateur stations as part of the self-training of the licensee in communication by wireless telegraphy.

Fig. 1. The Schedule of frequency bands, powers, etc, which, for the sake of convenience, appear in an identical format in both the Class A and Class B licences

Frequency bands in MHz	Status of allocations in the UK to: The Amateur Service	The Amateur Satellite Service	Maximum power Carrier PEP	Permitted types of transmission
1.810-1.850	Available to amateurs on a basis of non interference to other services.	No allocation.	9 dBW 15 dBW	Morse Telephony RTTY Data Facsimile SSTV
1.850-2.000		No allocation.		Morse Telephony Data Facsimile SSTV
3.500-3.800	Primary. Shared with other services	No allocation.	20 dBW 26 dBW	Morse Telephony RTTY Data Facsimile SSTV
7.000-7.100	Primary.	Primary.		
10.100-10.150	Secondary.	No allocation.		
14.000-14.250	Primary.	Primary.		
14.250-14.350		No allocation.		
18.068-18.168	Available to amateurs on a basis of non interference to other services. Antennas limited to horizontal polarisation, maximum gain 0 dB with respect to a half-wave dipole.	No allocation.	10 dBW –	Morse, AIA only
21.000-21.450	Primary.	Primary.	20 dBW 26 dBW	Morse Telephony RTTY Data Facsimile SSTV
24.890-24.990	Available to amateurs on a basis of non interference to other services. Antennas limited to horizontal polarisation, maximum gain 0 dB with respect to a half-wave dipole.	No allocation.	10 dBW –	Morse, AIA only
28.000-29.700	Primary.	Primary.	20 dBW 26 dBW	
70.025-70.500	Secondary basis until further notice. Subject to not causing interference to other services. Use of any frequency shall cease immediately on demand of a government official.	No allocation.	16 dBW 22 dBW	Morse Telephony RTTY Data Facsimile SSTV

Fig. 1. (cont'd.)

Frequency bands in MHz	Status of allocations in the UK to: The Amateur Service	The Amateur Satellite Service	Maximum power Carrier PEP	Permitted types of transmission
144.0–146.0*	Primary.	Primary.	20 dBW 26 dBW	
430.0–431.0	Secondary. This band is not available for use within the area bounded by: 53 N 02 E, 55 N 02 E, 53 N 03 W, and 55 N 03 W.			
431.0–432.0	Secondary. This band is not available for use: a) Within the area bounded by: 53 N 02 E, 55 N 02 E, 53 N 03 W, and 55 N 03 W. b) Within a 100 km radius of Charing Cross. 51 30'30"N 00 07'24" W.	No allocation.	10 dBW 16 dBW e.r.p. e.r.p.	Morse Telephony RTTY Data Facsimile SSTV Television
432.0–435.0		No allocation.		
435.0–438.0	Secondary.	Secondary.	20 dBW 26 dBW	
438.0–440.0		No allocation.		
1240–1260		No allocation.		
1260–1270	Secondary.	Secondary. Earth to Space only.		
1270–1325		No allocation.		
2310–2400				
2400–2450	Secondary. Users must accept interference from the ISM allocations in this band.	Secondary. Users must accept interference from the ISM allocations in this band.		
3400–3475		No allocation.		
5650–5670	Secondary.	Secondary. Earth to Space only.		
5670–5680				
5755–5765		No allocation.		

Fig. 1. (cont'd.)

Frequency bands in MHz	Status of allocations in the UK to: The Amateur Service	The Amateur Satellite Service	Maximum power Carrier PEP	Permitted types of transmission
5820-5830	Secondary. Users must accept interference from the ISM allocations in this band.		20 dBW 26 dBW	Morse Telephony RTTY Data Facsimile SSTV Television
5830-5850		Secondary Users must accept interference from the ISM allocations in this band. Space to Earth only.		
10000-10450	Secondary.	No allocation.		
10450-10500		Secondary.		
24000-24050	Primary. Users must accept interference from the ISM allocations in this band.	Primary. Users must accept interference from the ISM allocations in this band.		
24050-24250	Secondary. This band may only be used with the written consent of the Secretary of State. Users must accept interference from the ISM allocations in this band.	No allocation.		
47000-47200	Primary.	Primary.		
75500-76000				
142000-144000				
248000-250000				

Messages sent must be in plain language and concern matters of a personal nature in which the licensee or the person with whom communication has been established, has been directly involved.

Other forms of communication may be facsimile signals, radio teleprinter signals and visual images and signals (not in secret code or cypher) which form part of or relate to the transmission of message signals and images.

What is an Amateur Maritime Mobile Licence?

This is a special licence available to radio amateurs already holding a Class A licence which permits the operation of an amateur station on a vessel that is at sea or moored or anchored in any port, harbour or estuary. The normally allocated callsign is used together with the suffix /MM which denotes maritime mobile operation whilst underway, or with the suffix /MA which denotes operation whilst the vessel is moored or anchored. There are special clauses within the licence concerned with 'maritime mobile' operation and the frequencies allowed for this although the general requirements for the Class A licence still apply. Information about the maritime mobile licence can be obtained from the Department of Trade and Industry radio branch at the address listed at the end of this book. Note, that special tests are carried out on the equipment installed on the vessel by a representative of the Department of Trade and Industry to ensure that it complies with the regulations concerned with safety and non-interference to other radio and navigational apparatus on the vessel. The above requirements may be changed in due course.

Can an Amateur Station be used in other ways?

The station may be used as part of the self-training of the licence holder for radio communication during disaster relief operations conducted by the British Red Cross Society, the St John Ambulance Brigade, the Emergency County Planning Officer or any part of the Police Force in the UK or during any exercise relating to such operations. Messages to be sent by any of these authorised bodies must be communicated only via one licensed radio amateur and another. (See the question relating to RAYNET, Radio Amateur Emergency Network.)

Do records of transmissions have to be kept?

Yes, a record must be kept in one book (not loose leaf) and known as 'the log book'. This must show the dates and times (g.m.t.) of operation of the station, the callsigns of all stations worked, the type of emission (modulation) used and the frequency of operation at the time. This includes any CQ or general calls for other stations. The same applies to maritime mobile operation for which a separate log book must be kept. A log book may be kept for portable or mobile use although entries may be made in the main station log and marked accordingly.

Can recorded messages be transmitted?

Yes, but only those addressed to the station may be recorded and re-transmitted in accordance with the terms in the licence and provided re-transmission is intended for reception by the originating station only and that the callsign of that station is not included in the re-transmission.

Modulation is prohibited by means of recordings, other than as above, except special recordings of sinusoidal tones or tones within the audio frequency spectrum. Records or tape recordings of the type intended for entertainment purposes (e.g. music) may not be transmitted.

What is meant by a suffix added to a callsign?

A suffix is added to a callsign to denote operation in a place or circumstance different to that set down in the licence. For example, for temporary premises the suffix /A is used. In a vessel not on the sea, i.e. on inland waterways, or in a vehicle on the road the suffix /M (mobile) is used. For portable or pedestrian operation the suffix /P is used. Operation from aircraft is prohibited.

Are amateur transmitting stations subject to inspection?

The station, the licence and the log book shall be available for inspection at all reasonable times by a person acting under the authority of the Secretary of State.

Are certain forms of message prohibited?

Yes. The licence includes an important clause relating to the fact that a licensed station must not be used for business, advertising or propaganda purposes or for the sending of news or messages on behalf of or for the benefit of any social, political, religious or commercial organisation.

Are special phonetics used in sending callsigns etc. in speech?

The licence recommends the use of a uniform phonetic alphabet as contained in Appendix 16 of the Radio Regulations Geneva 1976. These phonetics should be used to emphasise the letters of the callsign and spelling in messages when transmission is in telephony (speech). The phonetic alphabet is as follows:

A	Alfa	N	November
B	Bravo	O	Oscar
C	Charlie	P	Papa
D	Delta	Q	Quebec
E	Echo	R	Romeo
F	Foxtrot	S	Sierra
G	Golf	T	Tango
H	Hotel	U	Uniform
I	India	V	Victor
J	Juliett	W	Whiskey
K	Kilo	X	X-ray
L	Lima	Y	Yankee
M	Mike	Z	Zulu

3

AMATEUR RADIO TECHNOLOGY

What is there to learn?

From a study of the syllabus for the radio amateur's examination given in the previous chapter, the amount of knowledge required to become fully conversant with all the technicalities involved may appear somewhat formidable. Such knowledge will not be acquired in a few weeks just by reading textbooks, but textbooks will be necessary. Without any prior knowledge whatsoever, one will have to begin with the first basic principles of electricity because every function that applies to, or is a part of, apparatus used in radio communication involves electric current, indeed without electricity there would be no radio, television or electronics.

In the space of this chapter it will only be possible to provide very brief and simplified explanations of some of the basic principles, theory and circuitry commonly associated with amateur radio, enough perhaps to give some idea of the large number of closely related but different items that make up the subject as a whole. It has already been stressed that evening class study, with the help of appropriate textbooks, is the only sure way to acquire sufficient knowledge to pass the RAE. Details of approved textbooks specially compiled for the RAE syllabus will be found in the Bibliography. Useful books in the same series as this include *Questions and Answers on Electricity* and *Questions and Answers on Electronics*.

What is an electric current?

The understanding of electricity requires some knowledge of the characteristics of the electron and its companion units that make up the structure of atoms and molecules. All matter is composed of molecules and these are the smallest particles that retain the complete characteristic of the substance. Molecules are composed of atoms that are themselves made up of smaller particles. Atoms can be visualised as having a central *nucleus* around which rotate one or more *electrons*. In metal the various atoms are situated in close proximity but some of the electrons are free to move about through the substance and are therefore called 'free electrons'. If however, an electric potential is applied between two points on the metal the number of electrons moving from the negative to the positive point will be greater than those moving in the opposite direction.

This *forced* drift of electrons along the metal, which may be a copper wire, is called an *electron current*. The movement of electrons is in reality an interchange of electrons between atoms and toward the positive terminal of the potential, although by convention the flow of current is regarded as being from positive to negative. In material that is called an *insulator*, e.g. ceramic, glass, mica etc. the number of free electrons is practically zero so electric current cannot be made to flow. Because metals of different kinds can be used to convey electric current they are called *conductors*.

What are the units of measurement?

The quantity of electrons that can be made to move is called a *coulomb* (symbol Q) and the amount of current flowing is determined by the number of coulombs per second. The unit of current denoted by the symbol I is the *ampere* (A). The force required to produce a given strength of current through a conductor is called the *electromotive force* (e.m.f.) (symbol E) and its unit is the *volt* (V).

Various substances will conduct electricity to a greater or lesser amount. Metals such as copper and brass are good conductors whereas iron and carbon are poorer conductors and offer some *resistance* to the flow of current. Insulating substances offer very high or infinite resistance so little or no current will flow at all. Because of this a relationship called

Ohm's law exists not only between the amount of current that can be made to flow along a conductor and the electrical pressure required to cause the flow but also the *resistance* (symbol R) offered by the conductor.

What is Ohm's law?

The most simple electrical circuit that could be devised would consist of a battery to supply e.m.f. and a conductor having some resistance, connected across the battery as in Fig. 2a. The current I flowing through the resistance R (measured in ohms) will depend entirely upon the voltage E of the battery. Ohm's law (after Simon Ohm) states that the ratio of the voltage across the resistance to the current flowing through it, $\frac{E}{I}$, is constant and will always be equal to the value of the resistance in ohms. However, if two of the values are known the unknown can always be found. For example if $\frac{E}{I} = R$ then $\frac{E}{R} = I$ and $E = I \times R$. The simplest way of remembering Ohm's law is by means of Fig. 2b.

There is another factor associated with current, voltage and resistance and that is *power* (symbol P) which is dissipated in the form of heat due to the flow of current through a resistance and which is measured in *watts* (symbol W). This can be derived from $W = E \times I$, or $\frac{E^2}{R}$ or $I^2 \times R$. Equally an unknown can be obtained from two known quantities:

$$E = \frac{W}{I} \quad \text{or} \quad \sqrt{W \times R}, \quad I = \sqrt{\frac{W}{R}} \quad \text{or} \quad \frac{W}{E}$$

$$\text{and} \quad R = \frac{W}{I^2} \quad \text{or} \quad \frac{E^2}{W}.$$

Fig. 2c shows the relationship of all these formulae.

Does amateur radio require much mathematical knowledge?

For the radio amateur's examination nothing beyond a reasonable knowledge of arithmetic and the ability to memorise a

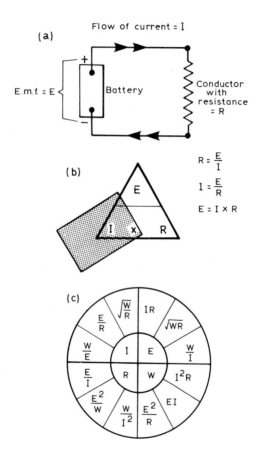

Fig. 2. (a) The flow of current through the resistance R will be in the direction of the arrows (see text). (b) An easy way to remember Ohm's law. Cover up the wanted quantity. The remaining two will provide the answer to this. (c) Relationships of formulae derived from Ohm's law

few 'end' equations and the signs and symbols associated with radio. Any arithmetic will be decimalised and large numbers will normally be expressed in index form, e.g. 1 000 000 as 10^6 with very small fractions like $\frac{1}{1\,000\,000}$ as 10^{-6} etc. Roots and square roots are used and one must understand simple graphs, how to interpret them and how to plot them. Ohm's law is one example of simple algebraic expressions of a number of quantities in which the symbols can be substituted for figures, for example if $\frac{E}{I} = R$ then find the value for R if $E = 20$ V and $I = 0.5$ A which comes to $\frac{20}{0.5} = 40$ ohms. Slightly more difficult would be:

$$\frac{1}{2\pi\sqrt{LC}}$$

which gives the resonant frequency of a tuned circuit when the inductance L is expressed in henries and the capacitance C is in farads. However, the inductance might well be given in microhenries (μH) and the capacitance in picofarads (pF), more likely fractions in practice, and the answer might be required in kilohertz (kHz), in which case the following formula would be easier to use. In this example the inductance L is 250 μH and the capacitance 160 pF.

$$f\,(\text{kHz}) = \frac{10^6}{2\pi\sqrt{LC}} = \frac{159\,155}{\sqrt{LC}}$$

which becomes

$$\frac{159\,155}{\sqrt{250 \times 160}} = \frac{159\,155}{200} = 795.775 \text{ kHz}$$

which by shifting the decimal point = 0.795 775 MHz or 795 775 Hz. The frequency is expressed in thousands of hertz (kHz), inductance in millionths of a henry (μH) and capacitance in millionths of a millionth of a farad (pF). Some of the quantities and units used in radio together with their respective symbols are shown in the tables.

Quantity	Symbol	Unit	Abbreviation
Time	t	Second	s or sec
Wavelength	λ (lambda)	Metre	m (sometimes mtr)
Frequency	f	Hertz	Hz
Electromotive force	E	Volt	V
Potential difference	V	Volt	V
Quantity of electric charge	Q (q)	Coulomb	C
Current	I	Ampere	A
Power	P	Watt	W
Inductance	L	Henry	H
Capacitance	C	Farad	F
Resistance	R	Ohm	Ω (omega)
Reactance (capacitive)	X_C	Ohm	Ω
Reactance (inductive)	X_L	Ohm	Ω
Impedance	Z	Ohm	Ω

What components are used in radio?

There are a very large number of different kinds of component each with its own particular application although many are derived from basic types. The major components are resistors, capacitors, inductors, valves and transistors plus certain other 'solid state' devices that have stemmed from the transistor. Capacitors are used for the brief storage of electricity that may be required by the function of some particular circuit. Inductors are used largely in circuits that are tuned to, or resonant at some particular frequency, or frequencies, as in the tuning circuits of a radio receiver or transmitter. Inductance is also

MULTIPLES AND SUBMULTIPLES OF UNITS

Prefix	Abbreviation	Multiplier
tera	T	10^{12}
giga*	G	10^{9}
mega*	M	10^{6}
Milo*	k	10^{3}
hecto	h	10^{2}
deka	da	10
deci*	d	10^{-1}
centi*	c	10^{-2}
milli*	m	10^{-3}
micro*	μ	10^{-6}
nano*	n	10^{-9}
pico*	p	10^{-12}
femto	f	10^{-15}
atto	a	10^{-18}

* Most commonly used in radio

used in transformers which are employed for the transfer of alternating voltage from one circuit to another without direct connection, as the mains transformer is used in a mains powered radio set. A valve, as its name implies, is a component used to control the flow of electric current but it can be made to amplify small electric currents and is therefore employed in all forms of amplifier circuits. Valves are also used in oscillating circuits for the generation of signals at frequencies ranging from a few hertz to many millions of hertz (hertz is the unit of frequency describing the number of cycles of an alternating voltage or current which occur in one second, e.g. 1 hertz (symbol Hz) = 1 cycle/second or 50 cycles/second = 50 Hz, the mains electricity supply frequency in the UK).

Although valves are still used in receiver and transmitter circuits the transistor has now virtually replaced them and almost anything hitherto done with valves is accomplished much more efficiently with transistors. Applications of the transistor are however, similar to those of the valve, i.e. they can be used as amplifiers and oscillators over a wide range of frequencies.

How is resistance used?

Resistance (or the resistor) is a commonly used component in circuits and employed in a number of different ways but primarily to limit the flow of current in some part, or parts of a circuit. A resistance may be fixed or variable and resistors vary in size from small carbon based types, less than a centimetre long for very low current use, to large wire wound types for heavy current. Values from fractions of an ohm to many megohms (millions of ohms). Variable resistors are more usually referred to as potentiometers.

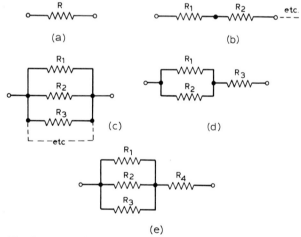

Fig. 3. (a) Single resistor. (b) resistors in series. (c) resistors in parallel. (d and e) resistors in series/parallel (see text)

In circuits resistors may be used singly or connected in series or parallel as shown in Fig. 3. Total values of resistors connected in series, parallel, or series/parallel can be calculated if the values of all the resistors are known, even though they may all have different values. The total value (R_t) of two, three or more resistors in series as in Fig. 3b is the value of all the values added together, e.g. $R_1 + R_2 + R_3$ etc. with R being the value of each resistor in ohms. For resistors in parallel the total value (R_t) is always less than that of the smallest value of the resistor in the circuit and with more than two resistors in parallel as in Fig. 3c is found by

$$\frac{1}{\frac{1}{R_1} + \frac{1}{R_2} + \frac{1}{R_3}} \text{ etc.}$$

With only two resistors in parallel, the total value is:

$$\frac{R_1 \times R_2}{R_1 + R_2}.$$

In a series/parallel circuit as in Fig. 3d the total value would be:

$$\frac{R_1 \times R_2}{R_1 + R_2} + R_3$$

but with more than two resistors in the parallel part of the circuit as in Fig. 3e would be:

$$\frac{1}{\frac{1}{R_1} + \frac{1}{R_2} + \frac{1}{R_3}} + R_4$$

How does a capacitor function?

A capacitor can be used in several ways but its basic function remains virtually the same and that is to act as a store for electricity for a brief period, which suggests that it can be charged with electricity. If two metal plates, separated by air

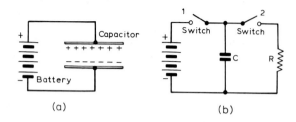
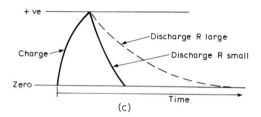
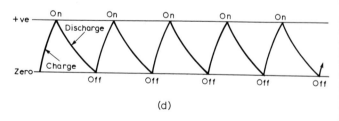

Fig. 4. (a) Formation of a capacitor. The two insulated metal plates will charge up to the potential of the battery when so connected (see text). (b) The rate of discharge of a capacitor can be controlled by a resistor as shown by (c). (d) Regular and continuous charge and discharge of a capacitor will produce a voltage waveform of finite shape and frequency

or other insulating material, are connected across a direct voltage supply (a battery) as in Fig. 4a then each plate will become charged, one positive and one negative. With the battery disconnected the metal plates which form the capacitor will remain charged. We can make use of this as shown in Fig. 5b in which the capacitor C is charged by closing switch 1 which is then restored to its off position leaving the capacitor

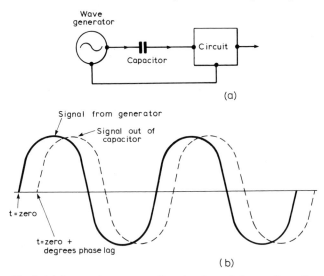

Fig. 5. (a) A generator or some other circuit supplying an alternating voltage can be coupled to another circuit by means of a capacitor although there will be a small time lag or phase shift introduced as depicted by (b)

charged. When switch 2 is closed the capacitor will discharge through the resistor R. Now it only takes a short time to charge a capacitor, i.e. for the charge to reach the potential of the supply, as shown by the graph in Fig. 4c. The capacitor could be discharged just as quickly as it was charged but if this is done through a resistor then the flow of current will be limited and it will take a longer time for the discharge to take place. The smaller the value of R the faster will be the rate of discharge and vice versa. If we could now operate the switches 1 and 2 in a regular but continuous sequence then a waveform would be produced as in Fig. 4d. By this means we have produced an alternating waveform at a given frequency. Instead of manual switching a valve or transistor can be made to do the work automatically and so produce a continuous wave.

Any continuously repeating wave such as a sine wave or square wave can be transferred from one circuit to another via a capacitor as shown in Fig. 5a and on the basis that the capacitor will charge up to one potential of the wave and at the rate at which that potential increases, and then discharge itself at the same rate and to the same level of potential into the next circuit. Because of the time taken to charge and discharge in this manner there is a small time lag between the original wave and the transferred wave as shown in Fig. 5b which is normally referred to as *phase difference*. Coupling one circuit to another by means of a capacitor is usually called a.c. coupling. Capacitors are also used in variable form in conjunction with inductance for tuned or resonant circuits in receivers and transmitters.

What is the difference between d.c. and a.c.?

Direct current (d.c.) is often used as a term to describe a steady voltage at a given level of potential since the current is also at a steady level. D.C. may be derived from a battery or accumulator or from an alternating voltage that has been rectified (changed to d.c.). Alternating current (a.c.) also refers to the voltage accompanying it and as its name suggests, alternates cyclically between a positive level through zero to a negative level. Let us return for the moment however, to d.c. and study Figs. 6a and 6b. The d.c. supply is represented by the battery which can be connected to a resistor (R) by the switch. With the switch open the potential across the resistor is zero as in Fig. 6b. When the switch is ON the potential at the top of the resistor will immediately become that of the battery and will remain so all

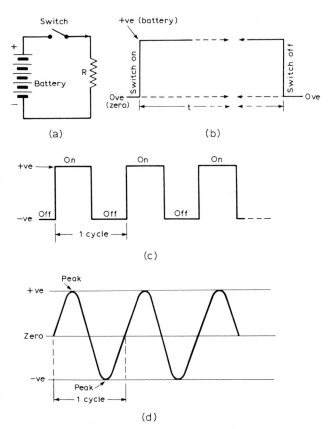

Fig. 6. (a) The function of direct voltage. With the switch closed the direct current (b) through the resistor R and the potential across it will be constant. (c) If the switch as in (a) is operated in a regular ON-OFF pattern an alternating current and potential will be formed into a repeating wave (see text). (d) An alternating voltage called a sine wave is formed when the voltage rises from zero to a positive peak, returns to zero and falls to a negative peak with a sinusoidal rate of change of amplitude

the time the switch is closed, but when the switch is turned OFF the potential across the resistor will fall to zero. However, if we were to turn the switch ON and OFF at a steady rate, say once each half-second, then the voltage across the resistor would change at the same rate as shown in Fig. 6c. Since the ON period is half a second and the OFF period half a second the full sycle is one second and it can be said that the cycles occur at a frequency of one cycle per second or 1 Hertz. In fact we have produced an alternating voltage. It is important to remember the switching on and off function at a steady repetition rate for this is the way in which certain types of valve and transistor circuits operate. The function itself is not only cyclic but the shape of the change of voltage graph (for this is what Fig. 6c really represents) resolves into a succession of square pulses. This can in fact be displayed on an oscilloscope and would appear almost exactly as shown in the diagram. It is known as a *square wave*.

All cyclically changing voltages of this nature are called *waveforms*. This leads now to alternating voltage with the most basic of all waveforms and one that rises and falls sinusoidally and is therefore said to be a sine waveform, or simply a sine wave. Because of the nature of its generation a sine wave does not rise from zero to a positive level only, but changes in polarity with respect to a zero point, i.e., the voltage rises from zero to a positive level then returns to zero and then 'falls' from zero to a negative level. The rate of change from one level to another is not instantaneous but gradual as in Fig. 6d.

Sine waves can be generated in various ways, the alternating voltage generator used in power stations being one of them with the common generating frequency of 50 cycles per second (hertz). Hertz comes from the name of the man who discovered radio waves and is used instead of cycles per second but the meaning is the same, i.e. 1 hertz equals one cycle per second. Sine waves can also be generated 'electronically' by valve and transistor circuits at a wide range of frequencies including the very much higher frequencies suitable for the transmission of radio signals.

What is the purpose of inductance?

Its main application in radio is in tuned or resonant circuits in receivers, transmitters and in oscillating circuits. An inductance (symbol L) consists of a coil of wire with the number of turns of wire and the diameter of the turns determining the value of the inductance which is measured in henries (abbreviation H). The function and all the characteristics associated with inductance are far too complex to deal with in a paragraph or so and any attempt to simplify would only leave the reader confused. But inductance can be applied in various ways. For example, there is *electromagnetic induction* which involves the use of a permanent magnet and a coil of wire (the inductance) for the purpose of generating electric current as illustrated in Fig. 7a. If we take an inductance L and pass through it a permanent magnet, the lines of force of the magnetic field will induce a small current into each turn of wire of the inductance. The total current adds up sufficiently to create a voltage across the coil and is indicated by the meter. Such a voltage is however, only of very short duration and the polarity is changed as the magnet passes through and out of the other end of the coil. If the magnet could be moved back and forth quickly and regularly then the voltage thus generated across the coil would be continuous but alternating like a sine wave. This particular application is the basis of the electric generator.

On the other hand by passing a voltage through an inductance a magnetic field can be created around it. Fig. 7b shows two inductances L_1 and L_2 close together, or technically, mutually coupled. If the switch is closed current from the battery will pass through L_1 and create a magnetic field around it which will be induced into L_2 so a potential more or less equal to that of the battery will be produced across L_2. However, such potential will be short lived as the fields collapse very quickly. If the switch could be operated repetitively and continuously then a repetitive potential would appear across L_2. We could of course replace the battery and switch with an alternating voltage in which case the repetitive currents in the coils and the magnetic fields would continue by themselves. A sine wave,

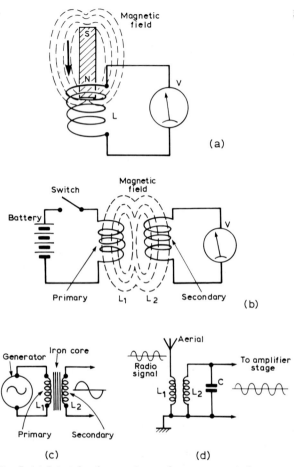

Fig. 7. (a) Principle of generating a voltage across an inductance. (b) Principle of inducing voltage into one coil (L_2) from another (L_1). (c) The transformer principle at low frequencies. (d) The transformer principle at high frequencies

or similar wave, could therefore be continuously transferred from one coil to the other. This is in fact the function of a *transformer* as shown in Fig. 7c, which consists of two coils wound around an iron core to improve the mutual coupling efficiency, but is a common application for isolating mains voltage supplies from mains operated radio and electronic equipment. The transformer application also makes it possible to step up or step down the voltage applied to the primary winding by adjusting the number of turns of wire on the secondary winding, e.g. a large number of secondary winding turns relative to those on the primary will step the voltage up and a small number will step the voltage down.

The transformer action is also used at radio frequencies for coupling from one circuit to another as in Fig. 7d. Here the iron core necessary for very low frequencies is not needed because the coupling efficiency at high frequencies is sufficient with air spaced coils although some improvement in this respect is often made with special iron dust cores between the coils. Coils operating at high frequencies can also be tuned or made resonant at a specific frequency by means of a capacitor C connected across the coil as in Fig. 7d. The response is such that virtually all other frequencies except the one the circuit is tuned to will be rejected. By making the capacitor variable we can also make the tuned circuit variable over a wide range of frequencies or wavelengths.

What is the relationship between frequency and wavelength?

All audio and radio waves and signals are alternating in character and may consist of a fundamental only, or combination of the fundamental and a series of harmonics which occur at frequencies higher than the fundamental but are nevertheless directly related. All fundamental waves and signals have a *frequency* which is the cyclic rate of change from one polarity to another as already explained. The audio range is roughly from 10 Hz to well over 20 000 Hz, the sonic range above audio from about 30 000 to 100 000 Hz and the normally used radio range from about 100 000 through to millions of

Hz, or megahertz (abbr. MHz) and on up to thousands of millions of hertz (gigahertz, GHz).

There is a very definite relationship between frequency and wavelength and also the velocity at which a wave travels. Radio waves in space and electrical signals of all frequencies along

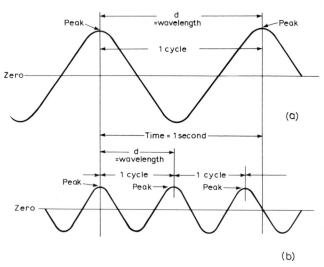

Fig. 8. (a) Wavelength is the distance (d) between two peaks of a waveform. Frequency is the number of complete cycles of the waveform in one second. (b) As the frequency increases the wavelength becomes shorter

wires travel at 300 000 000 metres (186 000 miles) per second and these velocities are constant. Sound waves in air travel at approximately 344 metres (1129 ft) per second. The wavelength is the distance (d) between the two peaks of a wave as shown in Fig. 8a but this distance decreases as the frequency becomes higher and as can be seen from Fig. 4b there are a little over two complete cycles in the time of one second as opposed to one cycle in the upper diagram. Because velocity is constant, wavelength varies only in relation to the frequency and vice versa. Very low frequencies have long wavelengths and very high frequencies have short wavelengths. The relationships between the three factors, namely velocity (v) in metres per second, the frequency (f) in hertz and the wavelength (λ) in metres are as follows

$$\lambda = \frac{v}{F} \quad f = \frac{v}{\lambda} \quad \text{and} \quad v = f \times \lambda$$

What is the difference between a valve and a transistor?

As far as general application is concerned, valves and transistors do similar jobs, for example either can be used as amplifiers and oscillators at very low and very high frequencies. The valve is a thermal emission or thermionic device, whereas the transistor is a semiconductor device. The most basic valve is the *diode* which contains one primary electrode called the anode and a heated cathode or a filament coated with tungsten which, when heated, gives off or emits electrons. There are other types of valve, e.g. the *triode* which has three electrodes. A diode and a simple circuit are shown in Fig. 9a. First a battery (L_t) is used to heat the coated filament or cathode. The positive terminal of another battery (H_t) of higher potential is connected to the anode with the negative terminal being connected to the filament or cathode. Electrons from the cathode will be attracted to the anode, now positively charged and which provides a conducting path for more electrons from the battery (H_t) as indicated by the arrows. Current will flow in one direction only, positive to negative and this will be shown by the meter. A diode is in fact a device for providing a conducting path in one direction only and is commonly used for the rectification of alternating voltage by allowing only the positive going half-cycles to pass through it and which become a crude form of direct voltage. A diode used for this purpose is commonly called a *rectifier*.

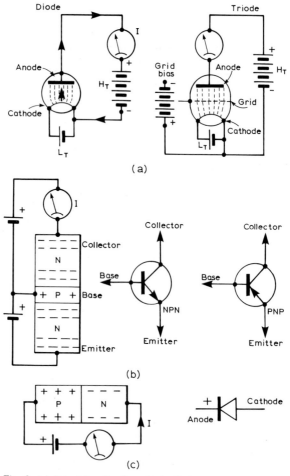

Fig. 9. (a) Principle of a thermionic diode and triode valve (see text). (b) Principle of the transistor (see text). (c) Principle of the semiconductor diode (see text)

How does a triode work?

A triode operates in much the same way as a diode in that it has the same form of electron emission from a heated cathode and also an anode with a positive potential applied to produce a flow of electrons and therefore a flow of current. The real difference is that the triode has a grid between the cathode and the anode. If a negative potential called grid bias is applied to the grid, the flow of electrons and therefore the flow of current between cathode and anode will be diminished, but if the grid is made positive then the electron flow will be increased and the current will be greater. A very small change in the grid potential results in a very large change in anode current. By placing a resistor in series with the anode, called the *anode load*, we can obtain a large change in the voltage developed across it by the change in current through it. Under this condition the valve can be made to *amplify*, i.e., a small change in potential at the grid will result in a large change in potential across the anode load. If the grid potential is derived from a specific waveform such as a sine wave, then the same waveform, but of course a larger version, will be present at the anode. There are other valves with more grids added (e.g. tetrodes and pentodes) to modify or control its functions, but the triode indicates basically how they all work.

How does a transistor work?

A transistor will perform functions similar to those of a valve but it stems from a semiconductor also called a *diode* and which is used in much the same way as a valve diode. However, the transistor operates on a principle quite different from that of a valve and consists of layers of semiconducting material as shown in Fig. 9b. This represents an npn-type transistor because the centre layer, the *base*, is of p-type material and the two outer layers, the *collector* and *emitter* are of n-type material. In a transistor known as a pnp-type the base is n-type material and the collector and emitter are of p-type material. These letters also give a clue to the main operating potential, i.e. npn transistors operate from a positive supply and pnp transistors from a negative supply. Otherwise their applications and operation generally are the same.

The operation of a transistor is not as simple to explain as that of a valve and to fully understand how a transistor works requires considerable knowledge of the various semiconductor materials and how they behave and which is also closely related with atomic theory. However, there is some similarity between transistors and valves and the following very simple but far from complete explanation is based on this. With a valve the control over electrons from the cathode is accomplished by means of a potential applied to the grid. In a transistor the emitter acts as a source of electrons which flow toward the base when this is biased positively with respect to the emitter. The collector corresponds roughly to the anode in a valve and in an npn transistor is supplied with a potential that is positive with respect to both the emitter and the base. Electrons arriving at the collector from the emitter are controlled by the nature and function of the base material and the external potential applied to it. This control potential largely determines the amount of current through the transistor from the collector to the emitter. A small change in potential at the base, or small change in current flowing therein, results in a large change of current into the collector. Like the valve the transistor is therefore able to amplify small signals applied to its input (base) and which emerge at the collector as identical but much larger signals.

A semiconductor diode (Fig. 9c) operates on a similar principle except that there is no controlling base material and current can flow in one direction only as it does through a valve diode. The symbols at the right of Fig. 9b and 9c show how the types of transistor and diode are indicated in circuit diagrams.

What are circuit diagrams?

A circuit diagram is a sort of shorthand way of showing the types of components in a circuit and how they are connected. Special component symbols, some of which are shown in

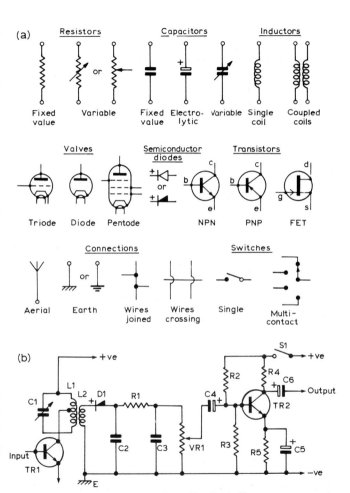

Fig. 10. (a) Some of the symbols used in circuit diagrams to represent components and connections. (b) Part of a typical circuit diagram (see text)

Fig. 10a, indicate what the component is while part of a circuit (Fig. 9b) shows how they are connected. It is usual in circuits to identify the components by a number and also to show their value, although actual values are not shown here. Component

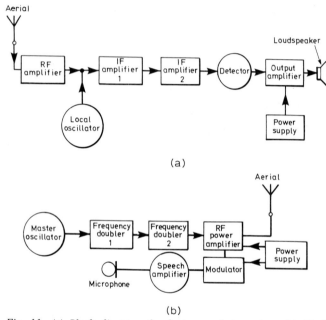

Fig. 11. (a) Block diagram of a basic superhet receiver. (b) Block diagram of a basic transmitter

numbers normally start from 1 across the circuit reading from left to right and this is also the normal way of reading a whole circuit. For example in Fig. 10b the circuit picks up a signal from a transistor TR_1 of a previous part of the whole circuit and can be explained in words in the following way: 'The transistor TR_1 is coupled via L_1 and L_2 (inductances) to a diode detector D_1 which connects with a simple RC filter network composed of R_1, C_2 and C_3. The output from this is connected to a variable resistor VR_1. From here the signal goes via C_4 to a transistor amplifier comprising TR_2, R_2, R_3, R_4, R_5, C_5 and C_6.' A description like this would also include a little more about the function of the circuit.

The ability to draw and read circuit diagrams is not difficult to acquire and proficiency comes largely through practice. There is, however, another but rather more simple form of circuit diagram but which does not show components and wiring. This is the *block diagram*, so called because blocks are used to indicate whole parts of circuits. Fig. 11a is a block diagram of a simple superhet receiver. Signals from the aerial go first to an amplifier stage and from here are mixed with signals from a local oscillator. The resulting signal is then amplified by the intermediate frequency (i.f.) amplifiers and passed to a detector which demodulates, i.e., separates the modulation signals such as music or speech, from the radio frequency signals. The demodulated signals are then amplified to drive a loudspeaker. If one already understands how the individual circuits work, i.e., those contained within each block, then the function of the whole system becomes apparent.

The block diagram of Fig. 11b is for a typical transmitter which begins with a master oscillator that determines the final but much higher transmitting frequency and which is therefore followed by two frequency multiplying stages, in this case a doubler (1) which explains itself and another doubler (2) the output signal from this being four times the frequency of the master oscillator. This signal is then amplified to deliver power to the transmitting aerial. As the transmitter is for speech modulation the signals from the microphone are first amplified and passed to the modulator stage (in effect a large power amplifier) which is used to vary the transmitted radio frequency power according to the frequencies of the speech.

What is modulation?

There are several different forms of modulation used in transmitting each involving a different technique and circuitry.

Modulation refers to the function of imposing speech or music, or any audio tonal signals onto the continuous high frequency carrier wave produced by a transmitter. Signals in the audio range are at relatively low frequencies and can be used to produce some form of variation in the carrier wave, for example, by causing its amplitude to change and which is known as *amplitude modulation*, or by causing the frequency of the carrier wave to vary and which is called *frequency modulation*. At the receiver the mode of modulation used at the transmitter is detected and then demodulation takes place which rejects the carrier wave but allows the original modulating signals through to the loudspeaker. There is also another and quite commonly used form of modulation known as *single sideband* (ssb) in which part of the carrier is suppressed before transmission and restored on reception. The function is very complex but it is a form of amplitude modulation and much more efficient in that power is saved and yet greater effective power is transmitted.

There is one other commonly used method of conveying information which is by using a radio carrier wave; this is called c.w., although in this case the carrier is not actually modulated by other signals but instead switched on and off, or 'keyed' in long or short bursts, for instance by a morse key, so that the dots and dashes of morse code can be transmitted. At the receiving end a high frequency signal is introduced to produce an audible tone which follows the on-off periods of the keyed carrier thus making the dots and dashes audible.

What abbreviations are used in radio?

The following are some of the more common ones.

a.c.	alternating current
a.f.	audio frequency
a.f.c.	automatic frequency control
a.g.c.	automatic gain control
a.m.	amplitude modulation
b.f.o.	beat frequency oscillator
c.w.	continuous wave

dB	decibel
d.c.	direct current
e.h.t.	extra high tension
f.e.t.	field effect transistor
f.m.	frequency modulation
h.t.	high tension
i.c.	integrated circuit
i.f.	intermediate frequency
l.t.	low tension
m.c.w.	modulated continuous wave
r.f.	radio frequency
r.m.s.	root mean square
s.s.b.	single side band
s.w.r.	standing wave ratio (also v.s.w.r.)
u.h.f.	ultra high frequency
v.h.f.	very high frequency.

4

EQUIPMENT FOR AN AMATEUR RADIO STATION

The range of equipment for amateur radio use is quite extensive and covers transmitters and/or transceivers, receivers and other items for home station use as well as transceivers for both mobile and hand-held operation. Other items include aerials (antennas), frequency measuring instruments, v.s.w.r. meters and many other accessories and components. Indeed many large dealers in amateur radio equipment may stock a thousand or more different items.

Can one build amateur radio equipment from separate components?

It is possible to construct an entire station including essential measuring instruments and other accessories from separate components but to do so would require extensive technical knowledge and a considerable amount of test equipment some of which could prove to be pretty expensive. Home construction is best reserved for auxiliary equipment for which circuits and constructional details will be frequently found in magazines devoted to amateur radio. It should be emphasised that home constructed transmitting equipment must comply with the technical requirements laid down in the amateur licence.

What about 'ready-to-build' kits?

At one time this was a means of saving money and easier from the point of view of construction for with a kit of parts everything is supplied. Kits are fine for small items of equipment but the cost may in any case prove greater than the complete ready-to-use article. Whilst a kit of components can provide the experience of construction, some previous expertise is desirable, particularly in assembly, the use of printed circuit boards and soldering. Great care is needed in this respect as integrated circuits (chips) as well as transistors and ultra-small components are commonly used.

What kinds of transmitters are used for amateur radio?

There are many transmitters to choose from that cover the various amateur radio bands but none that cover all the bands allocated by the licence. This applies to receivers as well. Transmitters and receivers and/or transceivers (T & R combined) come in two main categories. Firstly, there are the h.f. bands sets that mostly cover the frequency bands from 1.8 through to 28 MHz and which applies to transmitters, transceivers and receivers as well. One can therefore have a separate transmitter and a receiver to match, or the combined arrangement of a transceiver. Provision is made for the various modes of modulation, e.g. a.m. (amplitude), s.s.b. (single side band) and c.w. (continuous wave) for sending and/or receiving morse. Virtually all h.f. bands equipment has variable frequency tuning over the whole of each of the bands covered. Some have built-in crystal controlled frequency checking systems as well as digital readout of the frequency in use. Power output (transmitters only or transceivers) will be according to the price paid and price also determines various other facilities such as electronic frequency setting and recall via a built-in memory system. Extra r.f. power is obtainable with an external linear power amplifier driven from the main transmitter.

Equipment for the v.h.f. and u.h.f. bands is pretty much on the same lines but usually much smaller and with some limitation on power output. Extra r.f. power is possible with an external linear amplifier. Some v.h.f. and u.h.f. equipment has v.f.o. control, or similar, to permit operation on all frequencies within the band limits whilst others are channelised with up to

20 or so separate crystal-controlled frequencies but which, being very compact, are suitable for mobile operation. There are some transceivers even smaller, also channelised and designed specifically for 'hand-held operation or, as often more commonly described, as walkie-talkie sets. A large percentage of the communication receivers, transmitters and transceivers at present on the market are of Japanese manufacture, although American and British made equipment is available.

It is impossible within the confines of this book to even begin to describe the nature of the circuitry used in modern transistorised transmitting and receiving equipment and the reader will understand why from the following which is from the specification of a typical h.f. bands transceiver. 'It (the complete transceiver) contains 30 field effect transistors, 74 discrete transistors, 165 diodes and 5 valves.' A simple block diagram of this equipment alone would require the equivalent of about six pages of this book to reproduce and an abbreviated performance specification at least a whole page!

Are receivers for amateur radio purposes of a special kind?

The short answer is yes; the type normally used and which has to cover a wide range of frequencies is called a *communications receiver*. These are highly sensitive and normally transistorised receivers with special facilities not found on a normal broadcast band set. The frequency ranges on a standard type communications receiver will cover all the shortwave bands, including of course the various amateur bands, the total range being from about 1500 kHz (200 metres) down to at least 30 MHz (10 metres). For frequencies higher than this receivers are a little more specialised although today are usually integrated with a transmitter and called transceivers. A communications receiver will normally have switched frequency ranges with tuning for each range being spread across the full traverse of the tuning scale which will be calibrated in frequency and usually in wavelength as well. In addition specific narrow frequency bands, such as the amateur radio bands, can be 'bandspread' or expanded over a separate tuning scale so that individual frequencies are easier to find. This facilitates tuning because at very high frequencies this is very sharp. Such receivers are also equipped with separate controls for 'gain' at both radio and audio frequency, noise filters and special filters for reducing interference when stations are very close together or overlapping in frequency. Other facilities are provision for resolving different modes of modulation and which includes a beat frequency oscillator (b.f.o.) for receiving c.w. (morse code). Many modern v.h.f. transceivers have some but not all facilities similar to those of a communications receiver but they do have a means of providing a wide difference between transmitting and receiving frequencies for operation in conjunction with land and satellite repeater stations. Some transceivers are channelised, which means that operation on a fixed number of channels only is provided for. All transceivers normally and automatically transmit on the frequency band being used for reception and vice versa except for v.h.f. or u.h.f. repeater station operation.

What transmitting and receiving equipment is available?

More than enough to satisfy all requirements but the best way to get information about equipment is to write to the large retailers and/or manufacturers. The range is enormous and takes in hundreds of different receivers, transmitters, transceivers, station and auxiliary equipment, test equipment, aerials, rotators, morse keys including automatic types, slow scan video and teleprinter (RTTY) equipment etc, as well as thousands of components and solid state devices (transistors and integrated circuits). It would need an illustrated catalogue far larger than this book to cover the range available. There is little point in giving prices as they are constantly changing, as is the equipment available, because new models appear with great rapidity. A list of major suppliers appears at the end of the book which will be helpful and all will supply literature concerned with the products they handle.

There are of course many suppliers of amateur radio equipment and it is impossible to mention more than a few. However, advertisements for most suppliers and manufacturers will be found in amateur radio and electronics magazines. Before buying any equipment it is wise to study the performance specification etc. and consult with the suppliers as well as other radio amateurs as to the suitability of any item. The choice of equipment will in any case depend on particular requirements. For example, the Class B licence holder is restricted to operation on frequencies from 144 MHz and upwards i.e., the v.h.f. and u.h.f. bands. There is of course no limit to what one could spend but if cash is limited then some careful thought should be given to what can be spent on enough equipment to at least get on the air. For instance a small crystal controlled (channelised) 2-metre transceiver for 12 V operation could be used for both fixed station and mobile operation. For fixed station working it could be run from a float charged car battery or a suitable mains power supply. The only other requirement will be two aerials, one for the car and one for fixed operation, either of which would not be difficult to make. Later one could invest in a more ambitious transceiver for the fixed station and retain the 12 V set for mobile use only.

Even when one is able to acquire the full Class A licence some thought regarding aerials will be necessary. There is little point in purchasing an expensive all mode, all band, outfit if suitable aerials cannot be used due to lack of space or because planning permission cannot be obtained to erect both aerials and supporting masts.

Secondhand equipment is always available, usually by private buying, although some suppliers have secondhand transmitters, transceivers, receivers and auxiliary equipment for sale. It would be prudent to seek the help of someone more knowledgable if only to ensure that the equipment being purchased is at least in proper working order.

What auxiliary equipment is necessary and why?

Answering the second part of the question first, the terms of the transmitting licence state:

(1) A satisfactory method of frequency stabilisation shall be employed in the sending apparatus comprised in the station.
(2) Equipment shall be provided capable of verifying that the sending apparatus comprised in the station is operating with emissions within the authorised frequency bands.

What this really means is that one must (a) be able to check that transmission is within the allocated frequency band in use and that no appreciable energy is radiated outside the band. The licensee is required therefore to (b) use a satisfactory method of frequency control and (c) be able to ensure that transmissions do not contain unwanted frequencies, i.e. harmonics and spurious frequencies. When a station is inspected by officers authorised by the Secretary of State the licensee will be expected to demonstrate that he can comply with requirements (a), (b) and (c).

As a general rule a station requires a crystal reference source to comply with (a) and (b) above so that:

(i) with a crystal controlled transmitter an absorption device of suitable frequency range and accuracy is necessary to check that the desired harmonic of the crystal frequency is selected.
(ii) with a transmitter that is *not* crystal controlled a wavemeter based on a crystal oscillator is necessary.

Notes concerned with frequency checking equipment in amateur stations are attached to the licence and the subject is dealt with in classes for the radio amateurs' examination.

Most modern commercially made transmitters and transceivers are equipped with built-in crystal-controlled frequency checking systems and one may only require an absorption wavemeter to satisfy the terms of the licence. With home built equipment it is advisable to have both methods of frequency checking for all bands in use. Auxiliary items of this nature can be purchased from suppliers of amateur radio equipment.

What items of equipment are desirable but not absolutely essential?

Perhaps the most useful item for any amateur radio station is a multirange meter capable of measuring a.c. and d.c. voltage and current and resistance. This is a most useful instrument to have around when things go wrong but also invaluable for checking circuitry and the function of items of equipment that one might build. High-resistance meters should be used especially for transistor circuitry. There are plenty to choose from with voltage ranges covering zero to 1 000 V or higher, a.c. and d.c., and with current ranges from zero to 10 A and resistance from zero ohms to several megohms. Most modern multirange instruments have internal resistance of 20 000 ohms/volt or higher and are therefore more accurate than those with lower internal resistance because less current is taken from the circuit being tested.

A v.s.w.r. (voltage standing wave ratio) meter is essential particularly if much work with aerials is to be carried out. Some of the cheaper Japanese made meters also incorporate an r.f. power meter, or can be switched to read power and will be accurate enough for practical purposes. Many modern transmitters and transceivers, particularly those for v.h.f. operation, have built-in v.s.w.r./power meters.

An oscilloscope might be considered as something of a luxury but if a lot of constructional work is to be done there is no finer instrument to have in the workshop. New 'scopes are expensive but good secondhand models can often be obtained. However, the choice of an oscilloscope should not be made without some knowledge of its use and application. It would be worthwhile reading textbooks on oscilloscopes and their use before purchasing such an item. Aerials and rotators etc are dealt with in Chapter 5.

There are of course other accessories that one can buy, which include testing equipment of various kinds, other than multi-range meters and oscilloscopes. Among such items are digital reading frequency counters which are fairly expensive but can be home constructed, wavemeters and crystal controlled frequency calibration units, mains power supplies for 12 V d.c. operated transceivers etc., r.f. power amplifiers, or 'linears' as they are sometimes wrongly called but which enable one to run higher power from an otherwise low power transmitter or transceiver. Morse keys and microphones come under the heading of accessories although microphones are usually supplied with new transmitting equipment. Morse keys range from standard types to semi-automatic types and even some that can be programmed to send CQ and the station callsign automatically. The range is very large.

What about station safety precautions and station layout?

The actual layout of the station will depend entirely on the amount of space available and the equipment in use or that will eventually be acquired. It has been mentioned elsewhere that a station need occupy no more space than a small hi-fi system, possibly even less, but if the ambition is to set up a lot of equipment and also carry out constructional work then a spare room or large garden shed or similar outbuilding is ideal. Be warned, however, about the use of garden sheds even though they may be dry. In cold and wet weather considerable moisture can get in and this could cause severe damage to equipment. Be sure that the floor is well insulated as precaution against electric shock from equipment run from mains supply voltage. A wooden floor is satisfactory as long as it is dry, but concrete or earth could be dangerous.

If there is no shortage of space the best arrangement is a bench or table for the transmitting and receiving gear and essential items of auxiliary equipment with another bench and shelves for constructional work. Actual layout of equipment should be arranged so that one can operate seated in comfort without having to jump up and down to manipulate controls etc.

Safety of aerials and masts is dealt with in Chapter 5. Where any electrically operated equipment is used care must obviously be taken to ensure that mains wiring complies with safety standards and that all mains powered equipment is adequately fused. Wrongly fused electrical equipment and poor wiring are fire hazards and could also prove to be lethal. It is important that all mains powered equipment is properly earthed either via separate earth wires, or by the earthing line in the usual three cored mains cable. Good earthing is essential anyway with all transmitting and receiving equipment. When inspecting the interior of mains powered equipment always switch off first and pull out the plug. If tests must be made with the power on then have someone ready to switch off instantly in case of accidents.

5

AERIALS FOR AMATEUR RADIO TRANSMITTING STATIONS

What constitutes a transmitting aerial?

Aerials for transmitting have one very special feature, they are tuned to the wavelength in use, the technical term for this is *resonant*, and any aerial which is resonant at the frequency of operation may be a fraction of, or one or more whole wavelengths long; this is necessary if the aerial is to be efficient. It will be appreciated therefore that if one is operating on a low frequency (long wavelength) such as 2 MHz then one wavelength in physical as well as electrical terms is approximately 160 metres. An aerial of this length would require considerable space, probably far more than could be provided except by a very few amateurs fortunate enough to have acres of ground at their disposal. However, for this particular frequency, or rather the amateur band pertaining to it, 1.8 to 2 MHz, compromises can and normally have to be made. Although an aerial is also *naturally resonant* when it is a half-wavelength as well as a number of half-wavelengths long, it can also be made resonant by 'artificial' means at smaller fractions of a wavelength. Therefore one could use a quarter or one-eighth wavelength aerial which greatly reduces the total physical length. A quarter-wavelength for the 1.8 to 2 MHz band would be approximately 40 metres (about 132 feet). As the frequency increases and the wavelengths become shorter then the physical length of the required aerial also become shorter, for example a quarter-wave aerial for the 3.5 to 3.8 MHz band is about 10 metres long (approximately 32 feet). There is also another factor involved and in practice applicable mainly to aerials for the longer wavelengths called *harmonic resonance*. Most amateur radio bands are wavelength or frequency related, or as it is usually termed, harmonically related. For example 1.8 to 2 MHz is closely related to 3.5 to 3.8 MHz because the frequencies within the latter range are double the former. However, it is easier to think in terms of wavelength, in this case 160 and 80 metres respectively.

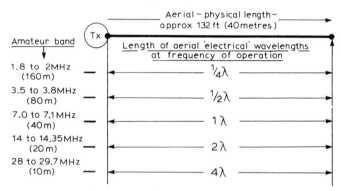

Fig. 12. Relationship between physical length and 'electrical' wavelength of an aerial with respect to five harmonically related amateur radio bands

An aerial one-quarter wavelength long for the 1.8 to 2 MHz band and tuned to resonance for that band will also be a resonant aerial one half-wavelength long for the 3.5 to 3.8 MHz band. It could therefore be used for either band. This harmonic relationship also applies directly as shown in Fig. 12 to three other amateur band allocations which in wavelength are 40, 20 and 10 metres. Some of the other amateur bands are not so perfectly related in this way but the use of compromise aerials is possible. However, aerials consisting of a single long wire, whether operated harmonically or otherwise, are commonly called long wire aerials and are suspended as high above the ground as possible. All resonant transmitting aerials are equally efficient for reception and therefore normally used for both purposes.

For shorter wavelengths such as 20 or 15 metres (14 and 21 MHz respectively), the physical and electrical half-wavelength is much smaller and at 20 metres for example is 10 metres long or about 32 feet. A half-wavelength now becomes the commonly used resonant length and at 10 metres (28 MHz) is 5 metres or about 16 feet. As we go higher in frequency so the half-wavelength becomes shorter and shorter and for the popular 145 to 146 MHz (2 metre) band is 1 metre or about 40 inches. This means that beam aerials consisting of several radiating elements can be used which are far more efficient than a single element aerial. The space required is of course much less, for example a 10 or 12 element beam aerial for the 2-metre band will be little more than 12 or 14 feet long and around 40 inches wide, an area of about 4 × 1 metres.

The kind of aerial(s) one uses will ultimately depend entirely on the amateur bands on which it is proposed to operate and the use of *all* bands could in fact call for several different aerials, so some compromise could be called for. A long-wire aerial with harmonic operation may well suffice for the bands 160, 80 and 40 metres. For 20, 15 and 10 metres a triple wavelength, or 'multi-band' aerial is feasible if one has the space to accommodate it. The very short wavelength bands (v.h.f.) such as 4 metres, 2 metres and 70 centimetres ideally require a separate aerial for each. For these bands beam aerials provide the greatest efficiency and it is not unusual to set them up one above the other at the top of a mast. Such aerials are of course highly directive and therefore have to be rotated, i.e. turned in the direction in which one requires to transmit and receive. However, for 2-metre band operation over 'local' distances of up to say 40 to 50 miles, or greater with suitable tropospheric or other conditions, a single element aerial may suffice.

Are the technicalities of aerials difficult to understand?

The subject of aerials taken as a whole is quite complex and very few amateurs make a study of a subject that offers considerable scope for experimental work. Only a few basic principles and information about popular types of aerials used by radio amateurs can be dealt with in this chapter, but as far as the radio amateur's examination is concerned a good working knowledge is required and this must also include the methods of feeding power to an aerial and matching it to the transmitter (and receiver) via a suitable feed line or cable usually called a *transmission line*. Some knowledge of radio wave propagation is also necessary. References to suitable textbooks are included at the end of the book.

What is radio wave propagation?

Radio waves travel in different directions depending on the frequency in use and their polarisation as they leave the transmitting aerial. For example at relatively low frequencies (long wavelengths) and from a vertical aerial, which means that the waves are vertically polarised, travel is close to the surface of the earth. This is commonly called *ground-wave* propagation. The distance for reliable communication is limited because the wave gradually loses power due to attenuation by the surface over which it travels.

At higher frequencies and with suitable aerials, usually horizontal and therefore the wave is horizontally polarised, the direction of travel is upward but at a relatively low angle away from the earth's surface. Such waves are reflected to earth again at some distant point by a region of the earth's upper atmosphere called the *ionosphere*. This is known as *sky-wave* propagation although the distance at which waves return to earth is governed to a large extent by the height and density of the ionised layers which vary with the intensity of ultraviolet radiation from the sun. Ionised layer variation also has considerable effect on radio waves at different frequencies to such an extent that at some frequencies the waves pass right through the layers or are only partially reflected. At other frequencies reflection may be total with the distance at which the wave returns to earth being dependent on layer height. The frequency, or it may be a number of frequencies, at which maximum reflection occurs in combination with the distance required to be covered, is called the *maximum usable frequency* or MUF. This may also be influenced by the time of the day. Generally speaking the greatest distances can be covered when the higher frequency bands such as 14 and 21 MHz are in use and also 28 MHz but at this frequency only when certain special ionospheric conditions due to sun-spot activity prevail. In order to make the most of ionospheric conditions the angle of the transmitted wave with respect to earth should be fairly low. The principle of radio wave reflection via the ionised layers is shown in Fig. 13. (See also Chapter 7)

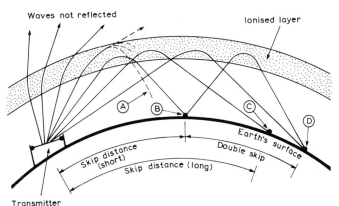

Fig. 13. Principle of long range radio communication by reflection of waves from ionised layers in the earth's upper atmosphere. Waves reaching (A) are only partially reflected. Waves arriving at (B) are fully reflected but the distance is short (short skip). Double skip occurs when waves arriving at (B) are re-reflected and returned to earth at a greater distance (D). Longer skip distances occur when the angle of reflection from the layer is large as for (C)

There comes a point however, when radio waves are no longer reflected by the ionised layers and this is when the frequency of transmission becomes very high (v.h.f.). Hence the wave propagation becomes what is sometimes called *space-wave* but more commonly referred to as *line-of-sight*, which means that the receiving aerial must be more or less within sight of the transmitting aerial. Provided that there are no large obstacles such as hills and high buildings in the path, the range to the horizon or farther will depend largely on the atmosphere near the earth's surface which under normal conditions causes radio waves at very high frequencies to follow a curved path. This keeps them nearer to the earth than would be the case had the line of travel been perfectly straight.

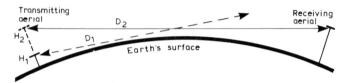

Fig. 14. When the transmitting aerial is at low height (H1) the wave path is D1 and signals do not reach the receiving aerial. Raising the height of the transmitting aerial to (H2) increases the range to beyond actual horizon distance by the path D2

Greatest horizon distance is achieved by maintaining the height of both the transmitting and receiving aerials as shown in Fig. 14. V.h.f. radio waves otherwise travel straight off the surface of the earth but at times weather conditions in the atmosphere at heights from a few thousand feet to a mile or so are responsible for bending them downward to some distant point. This is broadly known as *tropospheric* propagation and makes v.h.f. communication possible over much greater distances than can normally be achieved. Propagation in this respect improves as the frequency increases from about 50 MHz and is a phenomenon frequently prominent during spells of fine weather.

What is a transmission line?

Basically this is the line or cable that carries power from the transmitter to the aerial and/or signals from the aerial to the receiver. Any pair of wires that carry power from a source to a load, for example from a battery to a lamp, could be regarded as a transmission line. However, since we are dealing with power alternating at radio frequencies there are certain other factors to be considered such as the self-resistance, inductance and capacitance of the line which together create a characteristic known as the *impedance* (symbol Z_0) of the line. The transmitter output also has an impedance which must match the

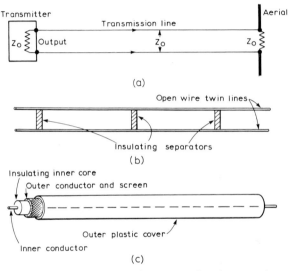

Fig. 15. (a) A transmission line connecting the transmitter must have the same impedance (Z_0) as that of the transmitter output terminal and also directly or by other means must match the impedance at the feed point to the aerial. (b) Construction of an open wire transmission line. (c) Construction of a coaxial transmission line

natural impedance of the aerial. Since the power required for the aerial must come from the transmitter then the line or cable connecting the two must ideally have the same impedance (Fig. 15a). If this impedance requirement is not met, for example if if the cable impedance is different from that of the aerial, then a mis-match would exist between the cable and the aerial and because of this power to the aerial would be lost. In such a case a special matching device becomes necessary.

Transmission lines take two major forms which may be open wire as in Fig. 15b or coaxial as in Fig. 15c. Coaxial cables perform in the same way as an open line but allow much lower impedance to be obtained, e.g. in the region 50 to 100 ohms, whereas the impedance range with open lines is from about 300 ohms upwards.

How do long wire aerials operate?

First we must consider what is called the voltage and current distribution on a radiating aerial a half-wavelength long. Such an aerial may be fed with power from the transmitter and assuming the transfer of this power is properly made then the current and voltage distribution along the aerial will be as shown in Fig. 16a. This diagram represents an 'instantaneous' picture and is really a graph which shows that when the current is maximum at some given point the voltage at the same point will be minimum and vice versa. This condition changes however, at the frequency at which power is supplied from the transmitter and current flows first to one end of the aerial and then back again during one cycle of the frequency. However, if another half-wave is added, current will flow along both but with a change in polarity as shown in Fig. 16b. The aerial is now two half-wavelengths long or one wavelength. We can therefore make an aerial consisting of a number of half-wavelength elements joined end to end as in Fig. 16c. It is also possible to use an aerial that is fractionally shorter or longer than a half-wavelength by tuning it so that the voltage and current distribution is equal to that of the single half-wavelength.

What is a radiation pattern?

This is really a graph that shows the strength of radiation from an aerial in all directions around it, e.g. in the horizontal plane through $360°$ and/or in vertical planes from zero to 180 degrees.

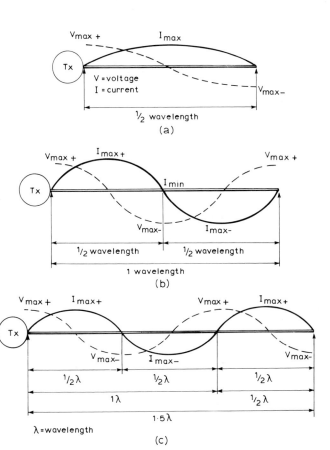

Fig. 16. (a) Instantaneous current and voltage distribution on an aerial half a wavelength long. (b) When an aerial is two half-wavelengths long or one whole wavelength, the current and voltage distribution is extended but becomes reversed in polarity. (c) Current and voltage distribution in an aerial 1½ wavelengths long. Note reversal of polarity in each successive half wavelength which follows this pattern as the number of half-waves becomes greater

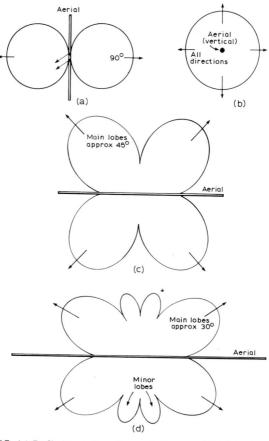

Fig. 17. (a) Radiation pattern from a half-wave aerial (dipole) is maximum at right angles to the line of the aerial. (b) When a dipole is turned vertically, radiation is equal in all directions around it. (c) Radiation pattern of a full wavelength aerial. (d) Radiation pattern from an aerial two wavelengths long

Such graphs can be plotted in Cartesian or polar co-ordinate form and are valuable to radio operators as they show exactly how an aerial behaves in terms of its directivity. The radiation pattern of a horizontal half-wave aerial, or dipole is shown in Fig. 17a and as can be seen, the shape is like a figure eight on its side with maximum level of radiation at 90° to the axis of the aerial. Radiation *in the line of the aerial* (from either end) is virtually zero but as there are two main fields of radiation (main lobes) such an aerial is considered as being *bidirectional*. However, the dipole and similar aerials can be used vertically in which case radiation is then at the same level in all directions around through 360° as in Fig. 17b. An aerial operated in this mode is said to be *omnidirectional*. There are also aerials that radiate mainly in one direction only and these are referred to as *unidirectional* or beam aerials.

Long wire aerials consisting of two or more half-wavelengths have multiple main radiation fields or lobes. For example, the pattern shown in Fig. 17c is that of an aerial one wavelength long. Maximum radiation is in four directions each at an angle of about 45° to the line of the aerial. Fig. 17d shows the pattern of an aerial two wavelengths long and there are four main lobes at about 30° to the line of the aerial and also four minor lobes at more or less right angles. It is important to consider radiation patterns, especially those of multiple half-wave aerials, with regard to the direction(s) in which one wishes to transmit with the greatest amount of radiation. For instance a full wave aerial with its pattern as in Fig. 17c would radiate with maximum efficiency in the directions NW, NE, SE and SW with the line of the aerial north to south. For maximum radiation in the directions N, E, S and W, the line of the aerial must be NE to SW or NW to SE.

How do beam aerials operate?

There are various forms of beam aerial so called because radiation is generally unidirectional, although there are special types that have bidirectional radiation. One of the most common forms of beam aerial is the *Yagi* named after its

Japanese inventor Dr. Yagi. This type of aerial is used mainly at very high frequencies although versions with two or three elements are used at frequencies as low as 14 MHz (20 metres) but these are rather larger being in the region of 10 metres (over 30 feet) wide and about 5 or 6 metres (about 20 feet) in

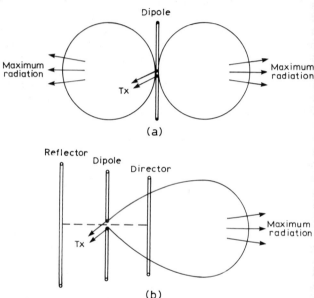

Fig. 18. (a) Radiation from a dipole is equal on both sides and maximum at 90 degrees. (b) By using a reflector and one or more directors radiation from a dipole can be concentrated with considerable increase in power in one direction only

length. Moreover, like all beam aerials of this nature they have to be rotatable to obtain maximum transmitted power in the direction required. Yagi aerials are also called *parasitic* arrays or beams and normally consist of a dipole main radiator which by itself has a radiation field that is bidirectional as shown in Fig. 18a. In order to concentrate the radiation from a dipole in one direction, it is necessary to introduce what are called *parasitic radiators*. A very simple form of Yagi type aerial may consist of the dipole and one parasitic element called a *reflector* which is precisely its function, to reflect radiation from one side of the dipole so that it becomes concentrated in what is referred to as the forward direction. However, it is possible to concentrate the radiation into a much narrower arc and thereby increase its strength by adding *directors* (also parasitic) in front of the dipole. A simple three element beam aerial consisting of a dipole with one reflector and one director is shown together with its radiation pattern in Fig. 18b. Parasitic elements are so called because the power they re-radiate is obtained from the driven or active element, i.e., the dipole. Aerials of this kind may consist of large numbers of elements which further narrows the radiation field but increases its strength more or less proportionally. Such aerials are more commonly used at very high frequencies because they are relatively small but capable of yielding very high gain. *Gain* means the increase in radiation obtained over that of a dipole and is usually quoted in decibels. For example a beam aerial with a gain of 12 dB (decibels) means that its radiated power in the forward direction is nearly 16 times greater than that of a dipole.

What is polarisation?

This refers to the orientation of the radiation field from an aerial and usually with respect to earth as the reference plane. Quite simply the polarisation of the wave from a horizontal aerial is *horizontal* and from a vertical aerial it is *vertical*. Wave polarisation takes other forms however, and from certain types of aerial can be elliptical, circular or slant, i.e. at some angle with respect to either vertical or horizontal.

What are standing waves and travelling waves?

These are terms that will be found in constant use when transmission lines and aerials are being discussed. Standing waves can be very much unwanted in transmission lines but are very important in all forms of aerials. The technicalities are rather complex and would be impossible to deal with in detail without large numbers of diagrams and virtually a whole chapter of this book.

A standing wave might be considered as the opposite of a travelling wave. However, if we take a transmission line as described in Fig. 15a and assume it to be perfectly matched to its source (transmitter) and load (aerial) then the wave along the line, composed of the current and voltage waves, will be constant, i.e., power will be flowing in one direction only toward the aerial where it will all be absorbed. This would be referred to as a *travelling wave*. If, however, the impedance of the line did not exactly match that of the aerial then some of the power would be returned down the line meeting power on the way up. The interference between the forward (to the aerial) and returned power would then cause a *standing wave* to be set up along the line and this in turn would cause the line to radiate. With an aerial it is important that a standing wave actually exists, otherwise little or no radiation would occur. Current flowing along an open wire of finite length has nowhere to go when it reaches the end, so it changes direction and returns along the wire. This process continues until the power generated by the flow of current backward and forwards becomes dissipated (a) by radiation and (b) in a very small proportion by loss due to the resistance of the wire.

It is very important as far as transmission lines are concerned that the standing wave level is very low and it is normally measured as a ratio to a unit of 1. The full name is *voltage standing wave ratio* or v.s.w.r. and the instrument used for checking it is a v.s.w.r. meter which measures the voltage associated with the wave.

What is an artificial aerial?

Before 1939 a special licence was issued that enabled those interested in amateur radio to build and experiment with transmitters providing they were not coupled to a radiating aerial but instead to an artificial aerial. This usually consisted of a tuned circuit with some form of resistive load, often an electric lamp, which dissipated the power from the transmitter and prevented virtually all of it from being radiated. Such a device is frequently used by radio amateurs today so that they can adjust and tune a transmitter without causing annoyance to others which would of course happen with the transmitter connected to an aerial. An artificial aerial can consist of little more than a pure resistance usually made of carbon and capable of dissipating the power fed into it. The value of the resistance is normally made equal to the impedance of the transmitter output terminal but is usually called a *dummy load*.

What is a field strength meter?

A very useful and easy to construct item of measuring equipment for checking the radiation from an aerial at close range. It also serves to indicate that the aerial is radiating to its fullest extent as the instrument incorporates a meter normally calibrated in equal divisions. For example if the meter is set to read full scale for normal transmission but shows a low reading then a fault is indicated, e.g. the transmitter power has fallen or there may be a bad connection somewhere. The field strength meter is very useful for comparing radiation between two different aerials but in this respect needs to be used carefully and not without knowledge of the conditions applying to such tests. A field strength meter usually consists of little more than a tuned circuit, a diode rectifier and low current meter with a short aerial attached.

What problems are there regarding the erection of aerials and masts?

From the purely 'engineering' point of view one could say there are none. Safety approved self-standing and/or guy wire supported masts and also short chimney-stack masts like those used for TV aerials can be purchased and erected by specialists.

Fig. 19. Typical bumper or roof mounting mobile aerial for v.h.f. operation (by courtesy of Lowe Electronics Limited)

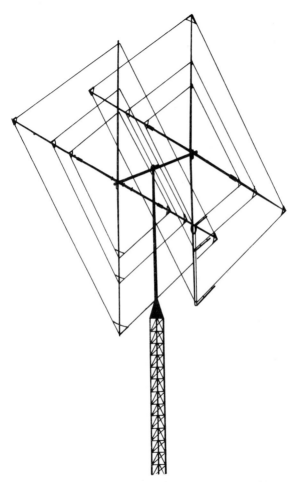

Fig. 20. Although of unusual appearance this Hy-Gain two element quad array operates on three wave bands, 10, 15 and 20 metres (by courtesy of Lowe Electronics Limited)

Provided one keeps the safety factors in mind a suitable mast could be made and set up as a DIY project. However, there is another and perhaps much more important point that must be considered. In the majority of residential areas one has to apply for 'planning permission' in order to add a small extension, to or make an alteration, to any property. Planning permission may almost certainly be needed for the erection of an aerial mast and aerial and possibly even for a small chimney stack mast with an otherwise unobtrusive aerial such as might be used for v.h.f. operation. It is up to the individual to make sure first as to whether such permission is required and will be granted. Apart from this there is the safety factor to consider as mentioned earlier and insurance cover against damage to property or persons. Aerials and masts have a habit of being blown down by strong winds, particularly if they are not too well secured.

Can transmitting aerials be purchased ready for use?

They most certainly can, but to most radio amateurs aerials provide a wide field for experiment and are always a topic for discussion. Almost all the common types of aerial for most of the amateur bands are available commercially with the possible exception of long wire aerials, but there are plenty of beam aerials and multi-band aerials available for bands ranging from 14 MHz (20 m) up to at least 430 MHz (70 centimetres). The full range also includes aerials for mobile use (1.8 MHz and all bands to 430 MHz) and a number of these can be adapted quite easily for portable operation as well. The cost of an aerial depends entirely on the type and the frequency band(s) it is to be used for. For example a v.h.f. mobile aerial as in Fig. 19 may cost as little as £10 whereas a large beam aerial for multi-band h.f. operation (Fig. 20) could cost up to £200 not including a mast and rotator which could easily add another £200. In view of the relatively high cost of some aerials it invariably pays to construct your own and various DIY designs for different amateur bands are frequently published in technical magazines.

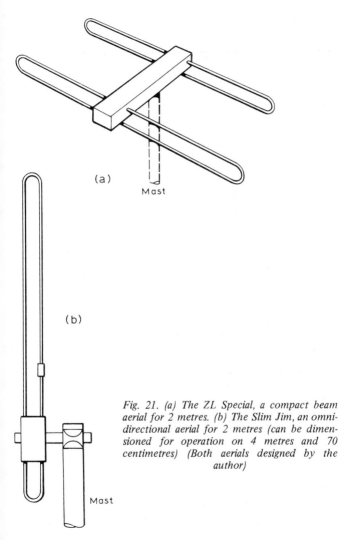

Fig. 21. (a) The ZL Special, a compact beam aerial for 2 metres. (b) The Slim Jim, an omnidirectional aerial for 2 metres (can be dimensioned for operation on 4 metres and 70 centimetres) (Both aerials designed by the author)

Is an aerial difficult to make?

A large multi-band aerial might be a difficult proposition though not beyond the bounds of possibility. Long wire aerials consist of little more than the necessary length of wire and some insulators and are therefore fairly easy to construct. Aerials for use at v.h.f. and u.h.f. offer considerable scope for the DIY enthusiast and designs are published in various handbooks (see list at the end of the book) and from time to time in amateur radio magazines. Two examples designed by the author are illustrated in Fig. 21. (a) is known as the ZL Special which is a compact two element (both driven) beam aerial for the 2-metre band. The 'Slim Jim' at (b) is widely used in practically every country in the world where 2-metre operation is popular. This is an extremely simple aerial to make, costs very little, and is suitable for all general coverage on this band as it is omnidirectional. It can also be dimensioned for operation on 4 metres or 70 centimetres.

Do transmitting aerials operate as receiving aerials?

It is virtually common practice to use the transmitting aerial for receiving as well and it is in fact more convenient to do this as opposed to using separate aerials. The performance of any aerial is reciprocal which means that it behaves in exactly the same way for receiving as it does for transmitting. For instance when a beam aerial is pointed in the right direction for maximum reception of a particular transmission it will automatically be in the right direction for transmitting. The changeover from transmit to receive is normally done automatically by a relay within either the transmitter or receiver.

What aerial accessories may be needed?

Rotators for beam aerials have been mentioned and the larger the aerial the more powerful must be the rotator. There are several types available, some being for very small lightweight aerials such as v.h.f. beams and some with high power for turning large arrays for the h.f. bands, these being very ruggedly constructed to carry the weight and turn the aerial in high wind. A small low power rotator will cost around £40 to £50 and a large one in the region of £120. Rotator units include a remote control box for turning the aerial from the operating position at the transmitter/receiver.

The only other major accessory used in conjunction with aerials is transmission line in the form of coaxial cable or possibly 300 ohm ribbon feeder. Open wire lines would almost certainly have to be constructed. It is important always to use good quality low-loss cable or line designed specifically for transmitting purposes and appropriate to the frequency band(s) used. Prices vary, but on average good quality coaxial cable costs around 60p per metre whilst 300 ohm ribbon feeder is around 10p per metre.

There are one or two special accessories which include aerial tuning units but items such as this may depend on the transmitter and aerial used and really comes under the general heading of station equipment.

6

OPERATING PROCEDURE, SIGNALS AND CODES

The operation of an amateur radio transmitting station is very like that of any professional radio communication service and most amateurs are quite proud of their ability in this respect, especially as wireless operators in the sending and reception of morse code. Radio amateurs have a code of operating discipline of their own and a generally quite unselfish attitude to all others using the often crowded frequencies. Frequency hogging, rudeness and other forms of selfish behaviour do occur, of course, but if one takes the percentage of this to the total amount of operating by the world's population of radio amateurs, it is very small.

Before dealing with questions about operating procedure it would be as well to deal with some of the codes, including the morse code, used by radio amateurs in both speech and c.w. (morse code) operation, particularly as certain codes are used in actual procedure, for example, one of the 'Q' codes, namely QRZ, means 'who is calling me'? This code would be used phonetically, or in morse code, when in doubt as to whether one was in fact being called by another station.

What is the 'Q' code?

This is an international code used in shipping and aircraft communications and by radio amateurs since the very early days of radio communication. It is understood and used by amateurs all over the world and consists of a number of three letter codes each beginning with the letter 'Q', the two remaining letters signifying the message in terms of both a question and an answer. For example, QSL means 'can you acknowledge receipt?' but also means in the reply 'I can acknowledge receipt' or 'I acknowledge receipt'. The Q code was devised mainly for use in communication by morse code to save time in sending an otherwise somewhat lengthy message. It is also frequently used in telephony (speech) transmission and has really become part of the jargon used by radio amateurs when talking to each other. For instance, you may hear something like this: 'Good evening old man, I copy you OK but there is some QSB and a little QRM. Shall we QSY'? Translated, the latter part simply means 'there is some fading on your signals and a little interference from another station, shall we try another frequency'? The other station may well quickly reply: 'QSL, QSY to S-zero' which means, 'I acknowledge receipt of your message, change frequency to 145 MHz'. The 'S-zero' is a channel number for the v.h.f. band 144 to 146 MHz (2-metre band) but more of v.h.f. and u.h.f. channel numbers later. The full list of Q codes will be found in amateur radio callbooks and other publications devoted to operating but the following short list contains some of the most commonly used.

QRK (q) What is the readability of my signals?
     (a) The readability of your signals is . . . .
QRL (q) Are you busy?
     (a) I am busy.
QRM (q) Are you being interfered with?
     (a) I am being interfered with.
QRN (q) Are you being troubled by static?
     (a) I am being troubled by static.
QRO (q) Shall I increase transmitter power?
     (a) Increase transmitter power.
QRP (q) Shall I decrease transmitter power?
     (a) Decrease transmitter power.
QSL (q) Can you acknowledge receipt (of message)?
     (a) I acknowledge receipt (of message).
QSO (q) Can you communicate with . . . .?
     (a) I can communicate with . . . .
QTH (q) What is your location?
     (a) My location is . . . .

It should be noted that these codes are frequently used in terms somewhat more loosely than those given above but with the general meaning of the code. For example: 'I had a QSO with G1XXX this morning' simply means: 'I had a contact with . . . .' etc., or 'I was in communication with . . . .' etc.

What other codes are used in amateur radio?

Aside from the Q code there is only one other and this is the 'RST' code which stems from readability, strength and tone and is as follows:

Readability

R1	Unreadable
R2	Barely readable
R3	Readable with difficulty
R4	Readable with practically no difficulty
R5	Perfectly readable

Strength (signal)

S1	Faint signals, barely perceptible
S2	Very weak signals
S3	Weak signals
S4	Fair signals
S5	Fairly good signals
S6	Good signals
S7	Moderately strong signals
S8	Strong signals
S9	Extremely strong signals

Tone (refers to quality of morse code signals)

T1	Extremely rough tone
T2	Very rough tone
T3	Rough tone
T4	Rather rough but better than T3
T5	Reasonably clean tone
T6	Clean tone
T7	Nearly d.c. tone i.e., a little mains hum audible
T8	Good d.c. tone, slight trace of hum
T9	Pure tone
T9X	Indicates clear crystal controlled transmission

The code would be sent thus: RST 559 or RST 599X etc. and is used mainly when transmissions are in morse code. Telephony operators use the S part of the code only to indicate signal strength e.g. S3, S5, S9 etc. but often give readability in full e.g. 'Your signals are S9, readability 5'.

The phonetic alphabet has already been mentioned in Chapter 2. In telephony phonetics are very frequently used particularly to ensure that callsign letters are not misread. For example, letters like X and S and P and B could easily be misread when signals are weak and there is interference on the channel. Radio amateurs invariably give callsigns phonetically anyway, especially when making contact with another station for the first time and the procedure would be as follows: 'This is Golf, figure two, Bravo Charlie X-ray calling Papa, Alpha, Figure 3, Alpha, Echo, November' which resolves into: This is G2BCX calling PA3AEN.

What other abbreviations do radio amateurs use?

Some of the more obvious are TX for transmitter, RX for receiver, DX for long distance, *rig,* which may describe the complete outfit, *ant,* which is used for aerial (antenna) and there are others already listed such as v.f.o., b.f.o., but which are general radio abbreviations anyway. Abbreviations other than those normally used in radio are often of everyday words and frequently used in order to reduce the length of messages sent in morse code. For instance, the following may or may not make sense:

GE OM UR SIGS HR RST 599 FB. QTH HR LONDON ES NAME IS FRED. TX HR 150W ES AND THREE EL BEAM. WX COLD BUT FINE. HW ME?

This form of jargon was at one time called 'hamese' and is not difficult to interpret but it does cut down the amount of sending in morse code very considerably. Some of the above are obvious but ES means *and* FB *fine business* and WX *weather.* Here is another message this time with the callsigns and other preamble that would be used when sending in morse code.

> VE G2BCX DE PA3AEN R.TKS UR CALL OM ES UR SIGS 589X WID QSB TO S2. WX ALSO COLD ES RAIN QTH VLAARDINGEN ES NAME IS WIM. PSE QSL. MNI TKS FB QSO ES 73. BEST DX. G2BCX de PA3AEN SK.

Sent in full the above message would have been something like this:

> R (received your transmission). VE (attention). G2BCX from PA3AEN. Thanks for your call old man and your signals are 589X (RST code) with QSB (fading) to S2. Weather also cold and raining. My location is Vlaardingen and my name is William. Please confirm this contact with a QSL card. Many thanks for a fine QSO (contact) and kind regards and best wishes (73). Wishing you more successful long distance contacts. G2BCX from PA3AEN SK (signing off code).

It is easy to see from this how the use of abbreviated words, Q code and procedure codes, etc. can convey an otherwise lengthy message in a short time. Incidentally the time-honoured figures 73 (means kind regards and best wishes) are always used at the end of contacts even in telephony operation. You will find the explanation of codes like VE, R and SK a bit further on when we deal with the morse code.

What is the morse code?

Most people know that it is a method of communication in which letters and numerals are sent as a series of dots and dashes, although in radio communication the short and long sounds are tonal, that is like short and long whistles. However, when one attempts to speak the code the terms dot and dash are too cumbersome and it is much easier to speak think in terms of dit (or di) and dah. For example the letter X is dash-dot-dot-dash but much easier to say or think of as dah-di-di-dah and which is the way to learn the code. You could memorise the following very quickly di-dah is A, dah-di-di-dit is B, dah-di-dah-dit is C. Do not attempt to learn morse the other way round, i.e., that A is di-dah etc., and the reason for this is that the sound has to be interpreted as a letter. So when you hear di-dah you will recognise the letter A. Note that punctuation marks are not used in the Post Office morse test, in fact the only one normally used by radio amateurs is the interrogation or question mark di-di-dah-dah-di-dit or IMI run together. Note also that the morse code as follows has been set down in order of code, phonetic and letter or numeral, etc. and the way to memorise it is to think di-dah is A, dah-di-di-dit is B and so on. (di is pronounced 'dee'.)

MORSE CODE

Code	Phonetic	Letter
•—	di-dah	A
—•••	dah-di-di-dit	B
—•—•	dah-di-dah-dit	C
—••	dah-di-dit	D
•	dit	E
••—•	di-di-dah-dit	F
——•	dah-dah-dit	G
••••	di-di-di-dit	H
••	di-dit	I
•———	di-dah-dah-dah	J
—•—	dah-di-dah	K
•—••	di-dah-di-dit	L
——	dah-dah	M
—•	dah-dit	N
———	dah-dah-dah	O
•——•	di-dah-dah-dit	P
——•—	dah-dah-di-dah	Q
•—•	di-dah-dit	R
•••	di-di-dit	S
—	dah	T
••—	di-di-dah	U
•••—	di-di-di-dah	V
•——	di-dah-dah	W
—••—	dah-di-di-dah	X
—•——	dah-di-dah-dah	Y
——••	dah-dah-di-dit	Z

· — — — —	di-dah-dah-dah-dah	1
· · — — —	di-di-dah-dah-dah	2
· · · — —	di-di-di-dah-dah	3
· · · · —	di-di-di-di-dah	4
· · · · ·	di-di-di-di-dit	5
— · · · ·	dah-di-di-di-dit	6
— — · · ·	dah-dah-di-di-dit	7
— — — · ·	dah-dah-dah-di-dit	8
— — — — ·	dah-dah-dah-dah-dit	9
— — — — —	dah-dah-dah-dah-dah	0

PUNCTUATION

· · — — · ·	di-di-dah-dah-di-dit	Question mark (?)
· — · — · —	di-dah-di-dah-di-dah	Full stop (.)
— — · · — —	dah-dah-di-di-dah-dah	Comma (,)

PROCEDURE SIGNALS

Code	Phonetic	Mark or sign
— · · — ·	dah-di-di-dah-dit	Oblique or stroke (/)
— · · · —	dah-di-di-di-dah	Break sign (=)
· — · — ·	di-dah-di-dah-dit	End of message (+ or AR)
· · · — · —	di-di-di-dah-di-dah	End of work (VA)
· — · · ·	di-dah-di-di-dit	Wait (AS)
— · — · —	dah-di-dah-di-dah	Preliminary call (CT)
· · · — ·	di-di-di-dah-dit	Also a preliminary call (VE)
· · · · · · · ·	di-di-di-di-di-di-di-dit	Error (eight dots)
— · —	dah-di-dah	Invitation to transmit

What is the best way to learn the morse code?

It should be possible to memorise at least four or five letters or numerals a day by mentally repeating them over and over for about fifteen minutes, i.e. take A, B, C, D the first day and memorise and so on. When you have memorised the alphabet it is time to begin practice with a morse key and buzzer and if possible to listen to real morse transmissions especially those put out by RSGB appointed amateur stations at slow speeds for learners. With at least 15 minutes a day practice, a sending and receiving speed of five to eight words a minute should be possible within a very few months but concentrate on receiving rather than sending. There are no short cuts to attaining speed, and learning letters by groups where the dot and dash symbols are similar does not help much either. Once you are able to copy at say 5 words or so per minute then always try to read something that is a little faster and which will speed up your reaction time. Even when you can read the requisite 12 words per minute continue and try for at least 15 or 16 before taking the 12 w.p.m. test. You will then be on the safe side. Spacing the dots and dashes and getting the correct duration for each which is what most beginners find difficult. One dash (dah) is equal to three dots (dits). The space between parts of the same letter is equal to one dot so with the letter A the timing would be di (space equals dit) dah. The space between two letters is equal to three dots so A, B would be di-dah (space di-di-dit) dah-di-di-dit. The space between two words is equal to about 5 dots.

Procedure signals, some of which resemble two letters together, are sent as one letter, e.g. the most used preliminary call is $\overline{\text{VE}}$. The bar above indicating a continuous code, i.e. no break between the letters and which would be di-di-di-dah-dit. Not many people know for instance that SOS is a long code and should be sent as di-di-di-dah-dah-dah-di-di-dit, and not as separate letters.

When groups of *numerals only* are being sent the figure 0 which is five dashes is often sent as a single dash to save time so if you copy 5T67T2 it is really 506702. If you are copying groups of letters and numerals mixed then any figure 0 (nought) is best written as $\emptyset$. So in a group like this, BC$\emptyset$ZO, the numeral will always be distinguishable from the letter O.

What is the significance of the callsign?

Amateur radio callsigns normally consist of the prefix which indicates country or nationality and a numeral and letters which are the station call and identification. For example, from the author's callsign G2BCX, the prefix G is for England and the 2BCX is the station call. A suffix is added when the station is operated under some other condition such as portable in which case /P is added to the callsign (G2BCX/P) with /M for mobile operation and /A for alternative address (see the author's

DUTCH AMATEUR RADIO STATION

PA3AEN

QRA LOC. CL∅2D

THE NETHERLANDS

W. PENNING_DE RUYTERSTR.3_3134-XN VLAARDINGEN

JA2 INQ

TOYOKAWA   JAPAN

Fig. 22

AMATEUR RADIO STATION JY1
CONFIRMS CONTACT WITH:

RADIO	DATE	GMT	MC	RST	2 WAY
			14		CW
			21		SSB
			28		

73

OP. HUSSEIN I
     P.O. BOX 1055
     AMMAN
     JORDAN

     PSE QSL TNX

Fig. 23

QSL card at the front of the book). Prefix lists for all countries are usually printed in log books and callbooks and are available from the Radio Society of Great Britain. Something about the callsigns used in Great Britain may however be of interest. For England the prefix is G, for Wales GW, for Scotland GM and each counts as a separate country. One can also get a good idea of how long British stations have been licensed by the callsigns. For instance, callsigns with G2 plus *two letters* are about the oldest in existence and some date back to the 1920s. Other pre-1939 callsigns are G5, G6, G4 and G3 plus *two letters*. Calls with G2 plus *three letters* were originally issued before 1939 with artificial aerial licences and for the full Class A licence in 1946. Calls with G3 plus *three letters* came into use from about 1946/7 and then came G8 plus *three letters* (Class B only) and G4 plus *three letters* (Class A licence). We now have G0 and G1 prefixes in use as well as G6 each plus *three letters*. There is also a G9 prefix (with three letters) but this form of callsign is only issued for special purpose licences.

What are QSL cards?

The code letters QSL mean acknowledgement of receipt (of a message) but radio amateurs use them to denote the cards they send to each other to confirm a two-way contact and which are known as QSL cards. It is a time honoured custom and many operators decorate the walls of their stations with these cards, especially those confirming DX contacts, for example, with Australia, New Zealand, South America etc. These cards are important in another way and that is as proof of contact with various countries for certificates of merit issued by the Radio Society of Great Britain the most notable being the DXCC award for achieving two-way contact with 100 different countries. The QSL card, usually postcard size, carries the station callsign and address etc., and has spaces for noting frequency, time, date, and information relative to the equipment and aerial etc. Some typical cards are shown in Fig. 22 from Holland (PA), Japan (JA), while Fig. 23 from Jordan (JY) indicates that not all radio amateurs are commoners!

Is a station log book a requirement of the transmitting licence?

It most definitely is but it should not be regarded as a chore, rather as a record of station activity as a whole including notes on experiments carried out, changes of equipment and aerials etc., as well as of all the contacts that are made. The licence requires a record of times and dates of all transmissions, with what stations and also of general calls. A log book can be made, suitably ruled for the various entries or can be purchased with pages already ruled and with columns titled for the required entries. They all have space for notes about experiments and tests etc., which although not required by the licence are nevertheless worth entering for reference.

How should a station be operated?

Certainly within the terms of the licence (Chapter 2) but also with common sense and with regard to others on the air. The procedure for c.w. (morse code) operation has been partly explained earlier but apart from the basic morse code letters and numerals, it is necessary to know the few procedure codes. The general call, or *CQ call,* is sent thus: CQ, CQ, CQ de G (callsign) and repeated say three times. Nowadays virtually all contacts or QSOs take place on the same frequency, usually the frequency of calling, unless a change of frequency is made by agreement. Your CQ call will be answered on the frequency you call on which means the answering station will 'net' onto your channel. The reply will be VE (attention call) G (callsign) repeated a few times and followed by, de ZL (his call) and the letter K which is the invitation for you to transmit.

If you are a good operator your message will be abbreviated in the style given earlier but never send too fast even if you are capable of doing so, because the other station operator may not be so proficient. If he sends too fast for you then request 'please QRS' (send slower). If you hear a station calling CQ then the procedure is much the same, i.e. net onto the frequency and call when he is finished.

Operation using telephony is similar and sometimes the calls may include spoken code letters for instance, 'This is G (callsign) calling CQ 20 (repeated a few times and ending with) 'and listening on the frequency K someone please'. The K (morse symbol dah-di-dah) is a signal code which means 'Go ahead', or 'Transmit please, I am listening'. Net operation is quite common particularly among stations local to each other and it helps to leave other frequencies clear. Net operating simply means that a number of stations use one frequency, usually by common consent and each takes it in turn to transmit.

Always listen for a short time on the frequency you intend transmitting on to make sure no other stations are using it. If by chance you do stumble on an existing QSO then find another channel. This is common courtesy that will be appreciated by others and by yourself should the same problem arise when you already happen to be in contact.

It has been mentioned that the v.h.f. and u.h.f. bands have channel numbers. For example the frequency 145 MHz in the

2-metre band is generally referred to as S zero (S0) but should really be R zero (R0) as it is an input channel for repeater station operation. These channel numbers are designated to the f.m. part of the band only (145 to 146 MHz), the remainder of the band from 144 to 145 MHz being devoted to other modes of transmission including s.s.b. and c.w. The channel numbers run from S zero upward with a change of number every 25 kHz, i.e. S1 will be 145.025 MHz, S2 is 145.050 MHz and so on. However a portion, or rather two portions, of the upper half of the band are devoted to repeater stations which require two channels, one being for stations working into the repeater and one for the re-transmission by the repeater of that station. Hence S zero is also R zero, a repeater input channel, with the output from the repeater being 600 kHz higher or on 145.600 MHz, which is R24 but has been used as S24, a simplex channel. Channels designated R are also known as duplex channels and those called S are simplex channels. Virtually all countries have repeater stations for the v.h.f. and u.h.f. bands and these are primarily intended for use by mobile stations to extend their working range. Satellite repeater stations operate on a similar basis and principle except that frequencies in two different bands are often used. Satellite repeater stations are known as OSCAR with a number to indicate which one. Details of their orbits and operation frequencies etc. can be obtained from the RSGB or the AMSAT organisation.

Are transmitting and receiving distances affected by frequency?

Yes and any one of the various amateur bands may be chosen according to the distance and coverage required. The greatest distance over which two-way contact is possible depends on the frequency band in use and atmospheric or tropospheric conditions as the case may be. Given the right conditions the 20, 14 and 10 metre bands are the best for round the world working. Beginning with the lowest frequency amateur band, i.e. 1.8–2 MHz (160 M), the general coverage and use is as follows.

1.8–2 MHz (160 m) This band is shared by radio amateurs and shipping and the interference level from ships and shore stations is very high especially near the coast. Long distance working is possible but this band is now mostly used for local contacts at ranges of up to about 50 miles. After dark the range is extended, often to 500 miles or so but results in this respect will depend on the aerial used and the level of interference from other services.

3.5–3.8 MHz (80 m) A very good band for working all over the UK, virtually the whole of Europe and Scandinavia and with the right conditions, across the Atlantic into Canada and the USA, especially on c.w. When conditions are very favourable it is possible, with a good aerial, to make contact with Australia and New Zealand as well as the Middle and Far East countries. Conditions vary with night and day but it must be emphasised that this band, as with some others, is now a 'shared band'. This means that other authorised services may use various frequencies within the band and must be given priority.

7–7.1 MHz (40 m) At one time a very good band for both European and transatlantic DX working but unfortunately this band is occupied by unauthorised broadcast stations for the greater part of the afternoon, evening and night periods. Even so some DX and/or local (UK) contacts are possible on c.w. and telephony.

10.1–10.15 MHz (30 m) One of the recently allocated new bands and for the time being may only be used for c.w., RTTY etc. (see The Schedule). This band has been little used but with the right ionospheric conditions is capable of long distance contacts around the world.

14–14.35 MHz (20 m) One of the most popular DX bands but at present and for perhaps another year or so (from late 1985) will provide only spasmodic contacts with the more distant countries, e.g. Far East and the America's etc.

18.068–18.168 MHz (16 m) Another recently allocated band but one which has to be shared with other services on a non-interference basis. Limited to c.w. only and the use of horizontally polarized aerials with unity gain, e.g. a half-wave dipole.

21–21.45 MHz (15 m) Comments about the 14 MHz (20 m) band apply to this also.

24.890–24.990 MHz (12.5 m) Same conditions as for the 16 m band above but a good band for DX when ionospheric conditions are suitable.

28–29.7 MHz (10 m) During the period of 'sun-spot maximum' which will not occur for another 4 to 6 years (from late 1985) this band is one of the most popular for world wide operation. The band will open for short periods even before the sun-spot maximum, often providing good North/South DX contacts.

70.025–70.7 MHz (4 m) This is a v.h.f. band and ranges are normally line of sight (see Chapter 5) but with good tropospheric conditions and/or Sporadic E during the summer months, contacts are possible up to 100 miles or more.

144–146 MHz (2 m) The most popular v.h.f. band with normally 'line-of-sight' ranges and one allocated to holders of a Class B licence. Good for local working up to 40 or 50 miles (fixed stations with good aerials). Mobile operation either direct or via repeater stations. Long distance contacts are possible using satellite relay, e.g. across the Atlantic etc., and considerable distances can be worked direct when the right tropospheric conditions prevail. Ranges of 500 miles are not uncommon.

432–440 MHz (70 cm) The first u.h.f. band and also allocated for use by Class B licence holders. Ranges under normal conditions are similar to those for 2 metres with interesting DX possibilities when conditions are suitable. Also used for mobile and repeater operation.

The remainder of the u.h.f. and s.h.f. bands are rather specialised and largely used for experimental work although some interesting long distance contacts have been made on the 1.36 Hz or 1300 MHz (23 cm) band. These bands may be used by Class B licence holders (see frequency allocations in Chapter 2).

7
MORE ABOUT RADIO WAVE PROPAGATION

What is the effect of the sun-spot cycle?

The effect of the 11-year sun-spot cycle was mentioned in the Preface because of the important part it plays in the propagation of radio waves at the higher frequencies between about 3 and 30 MHz. The h.f. amateur bands are located within this range and each can be individually effected to a very great extent by the radiation produced when sun-spots appear singly or in small groups on the surface of the sun.

It is the radiation from these spots, mostly ultraviolet, that cause particles in the regions high above the normal earth's atmosphere to become 'ionized'. It is these ionized regions, or layers, that reflect and/or refract radio waves in the frequency range mentioned above. Note: For the rest of this chapter the regions will be referred to as 'layers'.

Reflected or refracted waves are returned to earth which, being a reasonably good conductor, behaves as a reflector, or can cause refraction and therefore return a radio wave back to one of the ionized layers. There are in fact three main layers designated as E, F1 and F2. the lowest is the E layer at a height of about 100 km. However it will be easier to explain what is popularly known as 'skywave' propagation, the phenomena associated with it and the ionized layers, by continuing the format of question and answer.

What frequency range is effected by the E layer?

Mostly the range that forms the medium wave broadcast band and on up to about 3 MHz. With the exception of the 1.8–2 MHz amateur band, the E layer plays little or no part in the 'skywave' mode of propagation that applies to all amateur radio bands from 3.5 through to 30 MHz.

Which layer produces 'skywave' propagation of the amateur bands within the frequency range 3.5 to 30 MHz?

Mainly the F2 layer which is generally combined with the F1 layer and commonly known as 'the F layer'. Its height above the surface of the earth varies from about 200 km to 500 km, the average being about 350 km. But these are not the heights to the lower side of the layer which may be about 200 km in thickness. The 'height' figures apply to what is referred to as 'the virtual height' which is some way into the layer where a radio wave is turned round and returned to earth, either by direct reflection or refraction. The angle at which a wave is returned to earth is the same as the angle of incidence, i.e. the angle at which the wave meets the layer. The wave leaves the layer at the same angle.

How is all this effected by sun-spots?

The radiation produced by the appearance of a greater or smaller number of spots determines the density of ionization of the layer particles. The reflection or refraction of a wave depends on the density of ionization and if this is low the wave will have to penetrate some way into the layer before reflection or refraction can occur but this also depends on the frequency of the wave. As this is increased we reach a frequency where neither reflection or refraction can take place. This frequency is known as the frequency of penetration and the wave continues along its path through the layer and on into outer space as shown in Fig. 24.

Before this point is reached however and at a frequency a little lower than the frequency of penetration, we reach what is known as the 'critical frequency'. This is the frequency at which radio waves are directly reflected to earth when the angle of incidence of the wave at the ionospheric layer is 90 degrees.

Refraction at lower angles of incidence occurs at frequencies about 3.5 times the critical frequency. It is this condition that

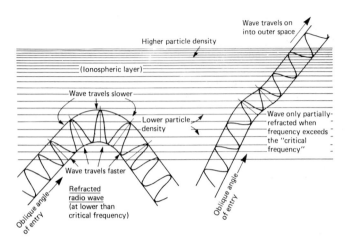

Fig. 24. The left hand portion of the diagram shows how a radio wave is 'bent' or refracted and leaves the ionized layer at an angle the same as the angle of entry. If the frequency of the wave exceeds the 'critical frequency', then the wave is only partially refracted and travels on into outer space (right hand portion of illustration)

determines the so called 'skip' distance. Remember the angle of exit from the layer is the same as the angle incidence which means that the wave will reach earth some distance from the point of transmission.

Since the wave is returned to the layer from earth, the process of refraction takes place again so the wave makes a series of 'hops' between the earth and the layer, be it the E or F layer. Again the reader is referred to Fig. 13, Chapter 5.

Why is this subject so important?

This question could be answered with another. How is it that radio waves can travel considerable distances around a spherical world? Radio waves otherwise normally travel in a straight line provided they do not meet a reflecting medium as described in the foregoing text.

The subject is more important than even many experienced radio amateurs imagine. The reason why radio waves (in the lower and higher frequency ranges) could traverse the spherical earth was not known even at the time when Marconi first succeeded in transmitting what were then called 'wireless waves', across the Atlantic ocean. It was some time later that the ionospheric layers were discovered and finally designated with the letters E and F by Professor E. V. Appleton in about 1930. There is in fact another region (not normally referred to as a layer) but known as the D region. More of this later.

The subject assumes greater importance at the present time because as mentioned in the Preface of this edition, the 11-year sun-spot cycle is near to its absolute minimum as can be seen from Fig. 25.

When is the F layer most effective?

Most long distance radio communication results from F layer refraction, the most applicable amateur bands being those between 3.5 and 30 MHz although reasonably reliable propagation as high as the 28 MHz band only prevails during the period of the 11 year sun-spot *maximum*.

This is not to say that during the period of the sun-spot minimum, contacts cannot be made at the higher frequencies, 21 or 28 MHz. These higher bands are open from time to time depending on the amount of sun-spot activity. Propagation may be mostly in a North/South direction providing a southerly path from the UK, to North and South Africa for example.

Is it possible to predict DX conditions?

Predictions are published by the RSGB in the monthly magazine *Radio Communication* and which stem from the Rutherford Appleton Laboratory at Chilton in Oxfordshire. These predictions are not 100% reliable but do provide some guidance as to which amateur bands are likely to be open for DX working.

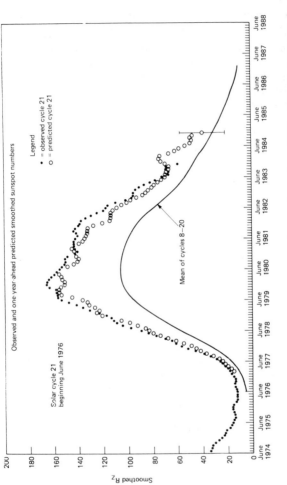

Fig. 25. Solar Cycle No 21 with the minimum starting at about June 1976, rising to maximum by about December 1979 then falling to minimum (last count approx. 60 Rz as at December 1983). The absolute minimum will not be reached until somewhere between about June 1986 or 1987. The next 'maximum' will not be reached until about five to six years later, between the years 1991 and 1993. These are predicted times and the closest that can be obtained. (Solar 11 year cycle chart by courtesy of the Belgian Royal Observatory).

Does the transmitting/receiving aerial play an important part in radio wave propagation when related to ionospheric conditions?

Most decidedly but few radio amateurs are able to put up aerials that would be most effective in this respect. The greatest limitation is usually the height of the aerial above ground.

What is the D region?

As already noted the D region is not strictly a 'layer' but a relatively dense part of the earth's lower atmosphere extending to a height of 60 to 90 km above the surface. In this part of the atmosphere the atoms are broken up into ions by sunlight but re-combine very quickly. The amount of ionization therefore depends on the amount of sunlight. However, this region has the effect of absorbing energy from a radio wave, particularly at frequencies from about 3 to 4 MHz and often as high as 7 MHz. The result is what is usually called a 'radio blackout' which can last for only a few minutes but can sometimes continue for a few hours.

What is 'sporadic E'?

Long distance radio communication is often possible during periods of sporadic E (generally abbreviated to Es). This refers to highly ionized 'clouds' of particles that are formed randomly and/or sporadically by the action of shear winds at a height of around 100 km, the same height as the normal E layer. Sporadic E occurs mostly during June and July although the author has known them to form in late May and early August. If the ionization is sufficiently dense, h.f. signals can be prevented from reaching the F layer so with Es the 'skip' distance will be reduced although the actual distance will depend on the vertical angle of maximum radiation from the transmitting aerial. Once formed sporadic E clouds usually drift in a southerly direction (from mid-latitudes) at a speed of about 100 km per hour. The clouds are sometimes small, in which case their usefulness as a propagation medium is short-lived. On the other hand an Es cloud can be very large, up to 200 km across and in this case, if the ionization is dense enough can propagate radio waves at frequencies up to 30 MHz or higher, e.g. over distances of 200 miles (320 km) or more. Communication may last an hour or so if the Es cloud is large enough. At v.h.f., e.g. 145 MHz (2 metres), Es propagation can extend communication distance to around 2000 miles (over 3000 km) providing that radiation from the antenna is at a reasonably low angle with respect to ground.

What is meant by MUF?

Reflection from an ionospheric layer can still be obtained when the frequency is increased beyond the critical, providing the wave enters the ionized layer at an oblique angle. This allows sufficient bending by refraction to take place and the wave to leave the layer for return to earth at an angle equal to the angle of entry. It is because of this and as already mentioned, that long distance transmissions can be made at frequencies some three to four times higher than the critical frequency and which are known as the 'maximum usable frequencies' (MUFs).

What is critical frequency and how is this determined?

Firstly the 'virtual' height of an ionospheric layer is measured by sending pulsed radio waves directly up to the layer, i.e. at 90 degrees to the earth's surface. The height is determined by the time taken for a pulsed radio wave to reach the layer and return directly to earth and based on the speed at which radio waves travel, i.e. 300 000 000 metres or 186 000 miles per second. As the frequency is increased during height measurements of this nature, we eventually come to a frequency at which the electron concentration is not dense enough to reflect the pulsed wave back to earth. The highest frequency that is returned and which applies to waves travelling 'vertically' to the layer, is known as the 'critical frequency'. Vertical incidence radio waves will penetrate the layer at all frequencies higher than the critical frequency and travel on into outer space.

What are the average critical frequencies for winter and summer during the periods of sun-spot minimum and maximum?

These are shown for any 24 hour period, therefore covering the hours of daylight and darkness, in the four sets of graphs, Fig. 26a, b, c and d. The maximum usable frequency (MUF) is between three and four times higher than the critical. To give an example, take the critical frequency for about 12 noon (all times are g.m.t.) for *winter sunspot minimum* (Fig. 26c). This is just above 6 MHz for the F2 layer (same for commonly called F layer). The MUF will therefore be around 18 to 24 MHz which offers the possibility of DX on the 18, 21 and 24 MHz amateur radio bands, as well as 7 and 14 MHz.

Can radio amateurs carry out ionospheric sounding?

No because this requires very special equipment and the use of pulsed radio waves over the frequency range from about 1 MHz to 25 MHz which is not allowed by the terms of the amateur radio licence. Ionospheric sounding is carried out in the UK by the Rutherford Laboratory at Chilton in Oxfordshire (World Data Centre). They have two ionospheric sounding stations, one at Slough in Middlesex which is linked directly to readout equipment at the Chilton laboratories and another located in the Hebrides at South Uist.

However the information obtained is made available to the RSGB and put to good use in compiling propagation predictions for radio amateur members. These predictions are published each month in the RSGB member magazine *Radio Communication*.

Can radio amateurs otherwise become involved in the study of radio propagation?

Yes. Sun-spot phenomena, as well as other solar activity, offers the possibility of serious study and quite recently a group has

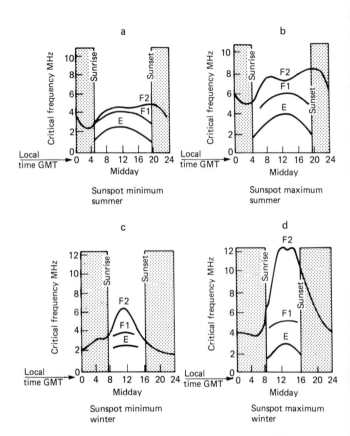

Fig. 26. Critical frequency curves covering a 24 hour period for (a) *sun-spot minimum-summer.* (b) *sun-spot maximum-summer* (c) *sun-spot minimum-winter and* (d) *sun-spot maximum-winter. Maximum Usable Frequency (MUF) is between three and four times higher than the critical frequency for the time of day and season during which the sun-spot count is either at minimum or maximum. (See example given in text)*

been formed for this purpose. Details from The London Solar Committee, 'Brindles', Mill Lane, Hooe, nr Battle, East Sussex, UK. There is also scope for the study of sporadic E events as well as 'tropospheric' conditions which frequency prevail and affect propagation in the v.h.f. regions. There is in fact a good deal more to amateur radio than simply making contact with other amateurs, near or far, as those who take up the hobby will discover by reading one or other of the magazines devoted to it. At the time of writing two-way communication tests on amateur radio frequencies (144 and 432 MHz) have been and are being carried out between radio amateurs on the ground and an amateur radio member of the crew of an American space shuttle in orbit.

Conclusion

If this book has been the means of providing the reader with at least a background to the hobby of amateur radio, then its object has been achieved. The author would like to thank numerous radio amateurs for suggestions regarding this new edition, the co-operation of the Department of Trade and Industry, The Rutherford Appleton Laboratory (World Data Centre) for supplying considerable information regarding ionospheric radio wave propagation which became the subject of a series of articles written by the author and published recently in *Practical Wireless*. (Extracts are reproduced in Chapter 7). Thanks are due also to the City and Guilds of London Institute for details of the Radio Amateurs Examination.

It would seem fitting to conclude with some short accounts of events in amateur radio that took place around 60 years ago when literally everything in the 'station' was home constructed, even individual components. The following, which has been taken from newspaper clippings, circa 1921/22, illustrates the enthusiasm that prevailed in those early days.

'2 MT' – MUSIC FROM THE MIST

A radio concert

'On Tuesday last a demonstration was given to the master and one hundred boys of the North Walthamstow Central School by the Walthamstow Amateur Radio Club....

'The concert, from 2 MT Writtle, was heard all over the hut; five valves and a loudspeaker were used.... During the interval the boys were given a short and simple explanation of the wireless wave and wavelength. The programme was concluded with an explanation of how to make a wireless set in thirty minutes for the price of two shillings.'

CONCERT BY WIRELESS

Audience of 5000 amateurs on a 700 metre wavelength

'It is estimated that over 5000 amateurs in this country were listening in to the Marconi wireless concert given last night on a 700 metres wavelength from Writtle near Chelmsford....

'Last night the results were much better than a previous attempt as the time had been altered to avoid clashing with the 'harmonics' of a Post Office centre which on the former occasion got the concert items mixed up with foreign despatches....'!

The above items are about the Marconi wireless station at Chelmsford in Essex, callsign 2 MT, which was the forerunner of the BBC in that it was one of the first to broadcast concerts for wireless as well as music enthusiasts.

The following are extracts from a newspaper account of wireless being used for the first time between air and ground not only to report on traffic conditions but to relay the results of the Derby (circa 1921/22).

'.... members of the Walthamstow Radio Club and others with wireless apparatus were fortunate enough to be at the phones during and after the Derby and had the opportunity of hearing from the airship R33 working in conjunction with Pulham, Croydon and other aerodromes around London.

'As the mighty silver streak glided over Essex little did the many thousands below realise how eagerly enthusiastic amateurs were waiting for all the latest news to fly on to their aerials; spoken direct from this marvellous glittering monster....' (the account concludes with) '.... we are told that flight and wireless telephony are in their infancy; what then will the future reveal?' *What indeed?*

The author is indebted to Mr. J. Pearce of Cantley, Norfolk, whose brother was one of those early radio amateurs, a member of the Walthamstow Radio Club and who was responsible for preserving the newspaper clippings from which the above quotes were taken.

APPENDIX: USEFUL INFORMATION

Books
A Guide to Amateur Radio, 19th ed, RSGB
ARRL Antenna Handbook, USA Publication available from RSGB
Amateur Radio Techniques, RSGB
Amateur Radio Operating Manual, 3rd ed, RSGB
Beginner's Guide to Amateur Radio, 2nd ed, F. Rayer and G. King, Newnes
Foundations of Wireless and Electronics, 10th ed, Scroggie and Amos, Newnes
How To Become a Radio Amateur, Department of Trade and Industry
756 Radio Amateurs Examination, City & Guilds of London Institute
HF Antennas For All Locations, Moxon, RSGB
Guide to Oscar Operation (Amateur radio satellites) AMSAT, RSGB
Two Metre Antenna Handbook, F. C. Judd, Newnes
Q & A CB Radio, F. C. Judd, Newnes

Magazines
Amateur Radio, Sovereign Publications (monthly)
Ham Radio, (USA Publication by subscription), RSGB
Practical Wireless, IPC Magazines Ltd (monthly)
QST, (USA Publication by subscription), RSGB
Radio Communication (Radcom), RSGB Members only

Electronics and Wireless World, IPC Business Press Ltd (monthly)
Datacom, Issued to members only of the British Amateur Teleprinter Group

Addresses

Newnes Books, Bridge House, 69 London Road, Twickenham, Middlesex TW1 3SB
RSGB Publications, (Sales) Lambda House, Cranborne Road, Potters Bar, Herts EN6 3JW
Department of Trade and Industry (Amateur Radio), Waterloo Bridge House, Waterloo Road, London SE1 8UR
City and Guilds of London Institute, 46 Britannia Street, London WC1X 9RG
British Amateur Radio Teleprinter Group (BARTG), Membership application to P. & J. Beedie, 'Ffynnonlas', Salem, Llandeilo, Wales SA19 6EW

Suppliers of amateur radio equipment

Amateur Electronics Ltd, 510-512 Alum Rock Road, Birmingham B8 3HX
Amateur Radio Exchange Ltd, 373 Uxbridge Road, Acton, London W3 9RN
Amcomm Services Ltd, 194 Northolt Road, South Harrow, Middlesex HA2 0EN
Arrow Electronics Ltd, 5 The Street, Hatfield Peverel, Chelmsford, Essex
Black Star Ltd, 4 Stephenson Road, St. Ives, Huntingdon, Cambs PE17 4EB
Bredhurst Electronics Ltd, High Street, Handcross, West Sussex RH17 6BW
Datong Electronics Ltd, Spence Mills, Mill Lane, Bramley, Leeds LS13 3HE
Dressler, 191 Francis Road, Leyton, London E10
Lowe Electronics, Chesterfield Road, Matlock, Derbyshire DE4 5LE

Maplin Electronic Supplies Ltd, PO Box 3, Rayleigh, Essex SS6 8LR

North London Communications Ltd, 400 Edgware Road, London W2

Radio Shack Ltd, 188 Broadhurst Gardens, London NW6 3AY

South Midlands Communications Ltd, SM House, Rumbridge Street, Totton, Southampton SO4 4DP

Thanet Electronics Ltd, 143 Reculver Road, Herne Bay, Kent CT6 6PD

Walters and Stanton Electronics, 18-20 Main Road, Hockley, Essex

Note: There are hundreds of suppliers of amateur radio equipment whose advertisements will be found in magazines devoted to the subject.

Radio interference

Hitherto, all cases of interference caused by amateur radio transmitters to other services, such as domestic radio, television, hi-fi systems, electronic organs and other 'electronic equipment' was dealt with by Post Office Radio Interference Branch. The complainant could obtain a form from any post office on which details of the interference were set down. Sending in the form resulted in a visit from an engineer who would then attempt to verify the cause of the interference. In cases where this was caused by amateur radio transmission the onus was usually put upon the radio amateur to effect some sort of cure. Until this was carried out to the satisfaction of both parties and the Post Office Radio Interference branch, the radio amateur could be ordered to remain 'off the air' until some sort of cure was found.

The situation has now changed as it was announced quite recently in Parliament that the Radio Investigation Service of the Dept. of Trade and Industry, would be devoting its efforts largely to those operating radio transmitters without a licence and those who, although holding a licence, resort to abuse of the terms and conditions of the licence. Secondly there will be far less time wasted on usually fruitless investigation into complaints of interference by owners of domestic radio, television, hi-fi and other 'radio or electronic' equipment used in the home. Much of this kind of equipment is prone to picking up interference of one kind or another due to lack of suitable internal circuit screening, poor quality design, inadequate maintenance, improper tuning and with TV the lack of a suitable high gain outside aerial. British Standard BS905 now provides minimum immunity standards for TV sets and which will be incorporated into legislation making it an offence to manufacture, sell or hire (or import) TV receivers which do not comply with this standard.

Owners of TV sets, radios and other domestic equipment prone to interference of 'any kind', including amateur radio transmissions, will be expected to deal with it themselves and with the assistance of the local dealer, or hire company or even possibly the manufacturer but aided by a booklet issued by the Department of Trade and Industry and available from Post Offices. However if you have to call in the DTI Radio Interference Service there will be a 'call-out' fee of £21. A 'log' of details of the interference such as times, dates and other relevant information must be provided. If this is not available and/or if the TV set operates from an indoor aerial, then the RIS will not co-operate. From 1987 the DIS will only investigate if your dealer/hirer declares that he cannot deal with the problem. The booklet mentioned above will no doubt provide further and more detailed information but it would appear that the onus is now fairly and squarely on the owners of domestic TV and radio equipment when it comes to the problem of interference not only from licenced radio amateurs but from other radiating sources, including CB radio.

INDEX

Abbreviations, 56
Aerials, 67–87
   masts, 81–84
   making, 86
   restrictions, 11–81
Alternating Current, 42
Amateur Radio
   activity, 4
   books and magazines, 14, 116
   clubs, 12
   development, 2
   examination for licence, 5, 16–19
   equipment, 58–61
   frequency bands (schedule) 25
   identification (callsigns) 3
   licences A and B, 23
   operating, 88–100
   origin, 1
   space required, 8
   technology, 31
   where to start, 14
Amplitude modulation, 56
AMSAT (Satellite operation), 12
Artificial aerial, 80
Auxilliary equipment, 63–64

Beam aerials, 77
Bibliography, 116
Block diagrams, 54
Building equipment (kits etc), 58

Callsigns, 30 and 95
Capacitors, 39–41
CB (Citizens band) Radio, 10–11
Circuit diagrams, 52

City & Guilds of London Institute, 16–17,
Clubs, 12
Components, 36
Conclusion, 113
Copy of licence, 22
Cost of licence, 23
Cost of station, 7–8
Critical frequency, 105–110

Dept of Trade & Industry (licences), 22
Development (Amateur Radio), 2
Distance (Transmit and receive), 100
DX predictions, 107

Electric current, 31
Equipment available, 61–63
Equipment Suppliers, 117
Examination (Radio Amateur Licence), 5, 16, 19–20

Field Strength Meter, 81
Frequency Modulation, 55
Frequency (wavelength), 47

General Information RAE, 17–18
Getting started, 14

Identification (callsigns), 3
Inductance, 35–40
Inspection of station, 29
Interference, 118
Ionospheric layers, 104–106

Kits for equipment, 59–60

Licence (cost), 6-8
Licence (schedule), 24-26
Log Books (station), 28-29

Mathematics required, 33
Maritime licence, 7, 28
Mobile operation, 27, 60
Modulation (amplitude, FM, SSB), 55
Morse code, 6, 20, 90
Morse code (CW operation), 92-93
Morse code tests, 21-22
MUF (Max Useable frequency), 110

National Society (RSGB) 12, 21, 22

Ohms Law, 33-34
Operation of station, 9, 27
Operating procedures, 88
 codes and practice, 89-91
Origin of amateur radio, 1

Phase shift, 41
Phonetics (telephony), 30
Polar diagrams (aerials), 76-78
Polarization (aerials), 79
Propagation (radio waves) 70, 104

Q codes, 88-89
QSL cards, 98

Radio amateurs' exam (RAE), 5, 16-20
RAE classes, 17
Radio wave propagation, 70, 104
Radiation patterns (aerials) 74-76
Radio Society of Great Britain, 12
RAYNET, 15
Recording (messages), 29

Receivers, 60
Resistance (resistors), 38-39
RST code, 90

Safety,
 station and equipment, 65
 aerials, 81
Shortwave listening, 10
Sideband modulation, 55
Slow morse transmissions, 22
Sporadic E (propagation), 109
Standing waves (transmission lines), 79
Station,
 cost of, 7-8
 inspection, 29
 layout, 65
Sunspot cycle (11 year), 104
Suppliers (equipment), 117
Surplus equipment, 7

Technology (amateur radio), 31
Test equipment, 63-64
Transceivers, 60-62
Transmitters, 59
Transmitting,
 aerials, 67-74
 licence, 16, 22-27
Transmission lines, 72-73
Transistors, 51
Travelling waves (aerials etc), 79
Triode (valves), 50

Valves, 49-50
Velocity (radio waves), 48
VSWR (voltage standing wave ratio), 79

Wavelength and frequency, 47

Yagi aerial, 77